Pollutants and Water Management: Resources, Strategies, and Scarcity

Pollutants and Water Management: Resources, Strategies, and Scarcity

Edited by:

Bhupendra Pushkar K.

Delve
PUBLISHING

www.delvepublishing.com

Pollutants and Water Management: Resources, Strategies, and Scarcity

Bhupendra Pushkar K.

Delve Publishing
224 Shoreacres Road
Burlington, ON L7L 2H2
Canada
www.delvepublishing.com
Email: orders@arclereducation.com

ISBN: 978-1-77469-107-6 (Hardcover)

ABOUT THE EDITOR

Dr. Bhupendra Pushkar was born in Mumbai, India. He persuaded his B.Sc. in Microbiology in 2000. He has obtained his M.Sc. degree in Biotechnology in 2002 from University of Mumbai. He pursued his further studies and completed his Doctorate degree in 2010. His area of research broadly includes environmental pollution remediation using biological science. He is currently working on heavy metal bioremediation, PAH bioremediation, plastic degradation, plant tissue culture etc. He is actively involved in the bioremediation research work and trying to develop system for river water cleaning. Since 2008 he is working as assistant professor at University of Mumbai, India. He has published various papers in national as well as international journals and currently guiding three research students. He is Co-Editor in Chief of a journal and reviewer in many journals.

TABLE OF CONTENTS

LIST OF FIGURES

clean drinking or drinking water for more than 1 billion people. This addition contains the consequences of water shortages and what it means to be without them

Figure 3.4. There are some rural regions in the whole world dealing with water shortages because they're just around the corner. Places deemed desert or remote areas cannot have sources in which humans are able to efficiently obtain water

Figure 3.5. The consumption of water necessary to support huge populations also increases as the world population. Water contributes to almost every area of a human life, although the creation of food is the majority. More than 70% of the world's water use is made up of agriculture. In addition, 2000–3000 Liters of water are necessary to generate adequate food for the daily consumption of calories for each individual (FAO)

Figure 3.6. While all of us depend on water to survive, we also contribute to increasing water shortages. A multitude of causes, including various human activities, are endangering the vitality of rivers, lakes and other bodies of freshwater. Pollution, climate change, industrial agriculture, unsustainable energy generation and population increase are all among them. The eventual outcome is an increase in water scarcity around the planet

Figure 3.7. A global catastrophe that is developing and is reaching over 785 million people all over the world is the Clean Water Crisis. Clean water is at the basis of the physical water shortage. Although clean freshwater is a critical resource for life, there is a lack of access to water for 1.1 billion people and a shortage of 2.7 billion of water every single month

Figure 3.8. Many water supplies that continue to promote ecosystems and feed an increasing number of people are stressed. Waterways, lakes, and rivers, dry or become too polluted to be used. There have been more than half the wetlands in the planet. Farming utilizes more water than any other source and, through inefficacy, wastes much of it. Climate change alters global weather and water patterns, causing droughts and shortages in some floods and areas elsewhere

Figure 3.9. There have been tremendous gains in the accessibility of clean water, with people without access to clean water falling from 1.1 billion in 2000 to 785 million in 2017. But the benefits of clean water may continue to be multiplied by improving sanitation and hygiene

Figure 3.10. In previously impacted areas, global warming is anticipated to increase the number of water strained places and exacerbate water stress. Subtropical locations such as Australia, the southern United States and countries of North Africa should get warmer and more severe droughts, although it is forecast to be more severe when precipitation occurs in these places. Climate scientists suggest that weather in tropical places also varies

Figure 3.11. The United Nations recognizes that Egypt will confront a significant water shortage in the next few years, requiring a more efficient management of the scant resources available. In order to avoid the existing waste, the irrigation systems should be modernized. If water shortages are not handled by those responsible, a country of about 80 million people will be in danger of running out of water in the next decades.

LIST OF ABBREVIATIONS

CAEWDP	Central Asian Energy-Water Development Program
CCSAs	Cross-Cutting Solutions Fields
DDT	Dichlorodiphenyltrichloroethane
EPA	Environmental Protection Agency
FAO	Food and Agriculture Organization of the United Nations
GAC	Granular Carbon Activated
GP	Global Practice Goals
HGMS	High-Gradient Magnetic Separation
HKH	Hindu Kush Area
IDA	Development Association
MDTF	Multi-Donor Trust Funds
MTBE	Methyl Tert-Butyl Ether
NBI	Natural-Based Infrastructure
NH_4^+	Ammonium
NOM	Natural Organic Matter
OECD	Organization for Economic Co-operation and Development
PAC	Powdered Activated Carbon
PCBs	Polychlorinated Biphenyls
POTW	Publicly Owned Treatment Works
SDGs	Sustainable Development Goals
SET	Sedimentation Erosion Table
UDSS	Urban Decision Support System
UNCHS	United Nations Centre for Human Settlements
UNEP	United Nations Environment Program
WASH	Water, Sanitation and Hygiene Issues
WHO	World Health Organization
WRM	Water Resource Management

WSG	World Solutions Group
WWDR	World Water Development Report

ABSTRACT

Water differentiates planet earth from all the other planets in the solar system. At the same time, as the global supply of freshwater is beyond enough to meet all contemporary and foreseeable water needs, its spatial and temporal allocations are not. There are numerous areas in which freshwater resources remain inadequate to fulfill domestic, socioeconomic growth, and environmental needs. In such areas, the lack of clean water to satisfy human consumption and sanitation needs is indeed a limitation on human health and productivity, and for this reason on economic growth as well as on the upkeep of a clean environment and holistic ecosystems. This book mentions that researchers must discover ways to take away these constraints. There are a couple of demanding situations in doing that, in particular given an unstable and unsure future climate, plus a rapidly developing population that is driving expanded socioeconomic improvement, globalization, and urbanization. How best to fulfill these needs requires research in various areas of water management. This book identifies the problems facing water managers today, and the research required to better inform individuals who attempt to develop a more sustainable and suitable future.

PREFACE

High-quality water is critical for the human lifestyle. However, human activities interfere in many ways with the natural water cycle and affect society. The constantly growing human population and its expectations concerning the standards of living demand proper exploitation of existing water resources. Despite its importance, water remains one of the most poorly managed resources on the earth. Upon contact with soil, the rainwater becomes surface water, and after soaking in, it is known as groundwater. For that reason, poor maintenance of surface water may cause major water problems, leading to infections of both humans and animals. This book suggests that the availability of safe water is also affected by global weather changes that cause shortages and overexploitation in a few locations, and flooding in other locations, which lead to decreased food protection and expansion of sickness.

Anthropogenic stress on the environment results in a decrease in water quality. There are many man-made causes of pollution that could contaminate water resources. With regard to their categorization, there are two classes which include, point and diffuse. Point source examples include commercial premises, cities, agricultural installations, manure garages, and landfills. They are easily identifiable and controlled compared to non-point resources, such as nitrates and insecticides, which mix with rainfall and cause heavy contamination over time.

The book also identifies two forms of water contamination: (1) Emergency contamination, often with immediately catastrophic impact, ensuing in loss of water fauna and many critical damages; (2) Long-term contamination, manifested by persisting natural pollution. It has a terrible net effect on the water ecosystem, ensuing in the absence of some species in the affected zones. Additionally, many infectious animal and human diseases are water-borne. They are normally transferred by way of ingestion of water contaminated with human or animal feces that comprise pathogenic bacteria, viruses, and parasites (protozoa, eggs of parasites). They'll live in the water for exceptional intervals of time, relying on water ecosystem factors. Monitoring of safety of water resources is primarily based on the dedication of parameters that suggest pollutants as a result of sewage, animal excrements, waste storage, animal manure, artificial fertilizers, and others.

The principal source of organic pollution within freshwater systems is the organic content derived from different human activities. This includes domestic and commercial sewage, wastes from agriculture and animal production, food processing facilities, and so on. Many toxic compounds are non-biodegradable or are degraded slowly so that they persist in the ecosystem; some are magnified within the food web; some may also cause cancer in people; or converted into carcinogens after they react with chlorine used to disinfect water; a few have a killing effect on fish and other aquatic organisms; some are nuisances, giving water and fish an offensive taste or smell. Acidification of inland waters by means of acidifying compounds of sulfur and nitrogen affects water quality and causes harm to aquatic ecosystems, particularly to fish.

Freshwater eutrophication is another global problem. Eutrophication (excessive proliferation of phytoplankton and filamentous algae resulting in greater turbidity, toxin production, diurnal variations in dissolved oxygen) is a result of water enrichment with phosphorous and nitrogen. Phosphorus discharges emerge predominantly from domestic and business effluents, but the ratio of agriculture isn't insignificant.

CHAPTER 1

WATER AND ITS USES

CONTENTS

1.1. INTRODUCTION

Water for a fact is an essential component in the world's existence. Taking the fact that even human beings are composed of water among other substances. It is considered a chemical substance given the fact that it reacts when mixed with certain substances. This substance is said to be transparent odorless, tasteless, inorganic and is also considered to be the main constituent of the world's hydrosphere. This also for the case of living organisms. Even though water is known to provide no calories or other important organic nutrients, it is very important in the life of living organisms (Zhao et al., 2017). As a matter of fact, individuals who have access to water can stay for a week without food but when one has no access to water, their chances of surviving is only three days. In those three days, the body will be using up the water that is present in it. H_2O is the chemical formula of water and single water molecule is made up of a oxygen atom and two hydrogen atoms. The two are connected by covalent bonds.

Figure 1.1. Drops of liquid water. Each of the water molecule contains a single oxygen atom and two hydrogen atoms. The chemical substance is essential for the existence of life in the world.

Source: Ars Technica.

Water is the name given to the liquid state of the H_2O component where it is at standard conditions of temperature and pressure. Upon precipitation, rain and aerosols formation take place in the form of fog. The solid state of water is ice. The clouds the sky are also composed of water. They normally contain suspended droplets of water and ice. Ice can be finely divided to form snow. Water vapor or stem is the gaseous state of water. 71% of the earth's surface is covered by water, which mostly consists of seas and oceans. However, there are those small portions that occur as ground water, glaciers and ice caps, and air, which composes of vapor and clouds. The three water states form a percentage of 1.7%, 1.7% and 0.001% respectively (Vyas & Nath, 2021). Water is always moving continuously in a cycle that it goes through evaporation, evapotranspiration, condensation, precipitation and runoff which normally moves to the sea and later again to evaporation.

Water plays a very important role in the world's economy given the fact that 70% of water goes to humans and agriculture. One can say that the entire 70% goes to humans given the fact that agriculture is essential for human survival. With agriculture, individuals can acquire the food that they need. Furthermore, fishing has become a very common component in both fresh water lakes and salty water lakes. Fish are a major source of food in most parts of the world. Apart from the fact that water is essential in providing food for human beings and other living organisms, it can also be used for transportation purposes for important commodities such as oil, manufactured products and natural gas. These products can be transported through seas, oceans, rivers and canals. Mostly, large quantities of utilized water can be reused for cooling, steaming, heating in the industries and in homes.

Figure 1.2. Water used for cooking. Water is essential for cooking in our homes. This is apart from the fact that it can be used in the transportation of substances all around the world.

Source: Pinterest.

Apart from the essential functions, water can also be used for the sake of recreational activities such as swimming, boat riding, diving, sport fishing, ice skating and skiing.

1.2. HISTORY OF WATER ON EARTH

In wanting to understand when water was on earth it is just important to grasp the concept that water is something that is continually lost through space and time. Water molecules are normally broken down to form free hydrogen atoms by the process of photolysis, which are sometimes seen to escape the earth's surface due to gravitational pull. Scientists claim that the loss of water from the surface was a very common thing when the earth was still younger. The reason of hydrogen escape from the earth's surface is its light weight such as helium. In the modern atmosphere, it is suggested that even the heavy components such as isotopic ratios of noble gases were subject to escape in the early atmosphere. A component described as xenon is one that the scientists claim can be used in estimating the amounts of water that have been lost through the atmosphere.

Researchers claim that any form of water on earth during the later times of accretion would have been lost due to the disruptions caused by the moon forming impact. This happened about 4.5 billion years ago and is the likely reason for the vaporized earth's crust. The upper mantle that then led to the formation of the vapor, which in doubt vaporized completely within 2000 years. Such occurrences might have been the reason for the presence of carbon dioxide in the atmosphere and the presence of hydrogen and water vapor. Initially, the surface temperature of the earth was 230°C and despite this, liquid water constituting of oceans were able to exist (van Koppen & Schreiner, 2014). The rise in temperature was caused by the increased atmospheric carbon dioxide. When cooling increased, there was are duction in the amounts of carbon dioxide on the earth's surface by the process of subduction which then dissolved in the ocean water and the levels were able to increase.

Scientists were able to discover geological evidence that shows the period in which liquid water existed on the earth's surface. Pillow basalt is a type of rock that was formed during the underwater eruption and can be very useful in explaining the time frame of the existence of water on the earth's surface. The rock was recovered from the Isua Greenstone Belt and by using it, scientists could proudly claim the existence of water on the earth's surface since 3.8 billion years.

Figure 1.3. Pillow basalts that are found in the ocean. These rocks are used to indicate the period of water existence on the earth's surface.

Source: EurekAlert.

Other rocks show that water has been in existence for over 4.28 billion years ago. If oceans however existed before the above mentioned time periods then, it means that there are other geological evidence that is yet to be discovered that will be useful for the process.

In more recent times, geologists have discovered that sufficient water might have been in existence to fill up our oceans since the time of planet formation. Apart from the use of rock, scientists can use minerals known as zircons, which are more resistant to weathering and other geological processes unlike with the case of rocks.

These rocks have proven very useful in helping individuals understand the conditions of the very early earth. The zircons have proven that water might have been in existence since 4.4 billion years which was soon after the formation of the earth (Afroz et al., 2014).

However, this evidence is some form of paradox given the fact that the temperature during the early days was cold enough to have frozen the water during that particular time. If it was true that the water existed during the time, the high carbon dioxide levels on the earth surface would prove that the earth during the early stages was not that different from the earth, as we know it in this period.

1.3. HYPOTHESES FOR THE ORIGINS OF EARTH'S WATER

1.3.1. Extra Planetary Sources

Unlike other materials within the terrestrial planet in the solar system, water is considered to have a low condensation temperature. During the early history of the solar system, the region described as the protoplanetary disk which is close to the sun was known to be very hot and thus one cannot say that water condensed in as the earth formed. The frost line was the name given to the boundary region where ice was able to form during the early stages of the solar system. This is located in the modern asteroid belt, which is about 3.1 astronomical units from the sun.

One speculation claims that earth accreted which means that it steadily grew via means of accumulation of icy planetesimals approximately 4.5 billion years ago, whilst it became 60 to 90% of its modern-day size. In this scenario, earth became capable to keep water in a few shapes in the course of accretion and fundamental effect events (Singh et al., 2021). This speculation is supported via way of means of similarities within the abundance and the isotope ratios of water among the oldest recognized carbonaceous chondrite meteorites and meteorites from Vesta, each of which originates from the Solar System's asteroid belt. It is likewise supported via research of osmium isotope ratios, which recommend that a massive amount of water became contained within the cloth that earth accreted early on. Measurements of the chemical composition of lunar samples accrued via Apollo 15 and 17 missions, in addition, assist this and suggest that water became already gift on earth earlier than the Moon became formed.

One hassle with this speculation is that the noble fuel line isotope ratios of earth's surroundings are extraordinary from the ones of its mantle, which indicates they had been shaped from extraordinary sources. To explain this observation, a so-called "overdue veneer" principle has been proposed wherein water changed bringing about a great deal later in earth's history, after the Moon-forming impact. However, the modern-day expertise of earth's formation permits for much less than 1% of earth's cloth accreting after the Moon shaped, implying that the cloth accreted later were very water-rich. Models of early Solar System dynamics have proven that icy asteroids might have been brought to the internal Solar System, which also includes the earth for the duration of this era if and whether Jupiter migrated toward the sun.

1.4. FACTS ABOUT WATER IN THE WORLD

Water is a substance that is composed of hydrogen and oxygen chemical component. It is known to be existing in solid, liquid and vapor states in nature. Generally, in the world, it is found almost everywhere in plentiful quantities. The substance has also found itself to be very essential among all compounds due to the various uses it is majorly associated with. Water is tasteless, odorless and most of the time clean water is colorless at room temperature. Many studies have reported that water has an imperative capability to dissolve in other substances in nature. Water can dissolve in other liquid substances to create other solutions or to reduce the concentration of chemical components in dissolvable liquids. To living organisms, water is essential for survival purposes. Water is believed to be very essential in life as it is termed as life itself. Aqueous solutions such as blood in animals and human is mostly composed of water. The consumption of water by the living creatures increases the concentrations of water in their blood and bodies. The digestive juices in animals and also humankind is also concentrated mainly with water.

Water exists on all the planets and on the moons, and goes even beyond the solar systems in the universe. In terms of appearance, water appears colorless when it is viewed or observed while in small quantities. On the other hand, in large quantities water has an observed intrinsic color that is light blue, which is said to be caused by minimal absorptions by the lights at wavelengths, which appear in red. The water compound is represented by a chemical structure (H_2O) that is physically, chemically as well as in its state known to have extremely complex properties. There are however long debates among scientists about the origin of the water. They have debated on the circumstance on whether the earth's water and even the other planets were formed as the earth was created or it would have come up later. Although from the biblical perspective, the waters were there since times immemorial. The Almightily had to separate the waters from the lands that had been in existence since time immemorial so that he would give space for land animals, plants and other structures existence. Various studies in the past have concluded that a majority of the earth's water found its origin from the rocks in the mountainous and areas that are forested from which the earth is located. Other researchers have argued that water encompasses everything thus its origin is a mystery. In the earth, water is not typical of many other substances such as the observable sight of ice cubes that may be floating in a glass of ice water substance may be seen as very unusual for other chemical entities while with water the conduct may appear to be very common. For most other chemical compounds, it has been proven through research that the solid-state would always be denser than the liquid state of a compound. This would mean that the solid-state compound would sink in the liquid state compound while immersed. It extremely special that ice on the other hand floats on water. This means that ice is less dense than water in liquid state (Singh et al., 2020).

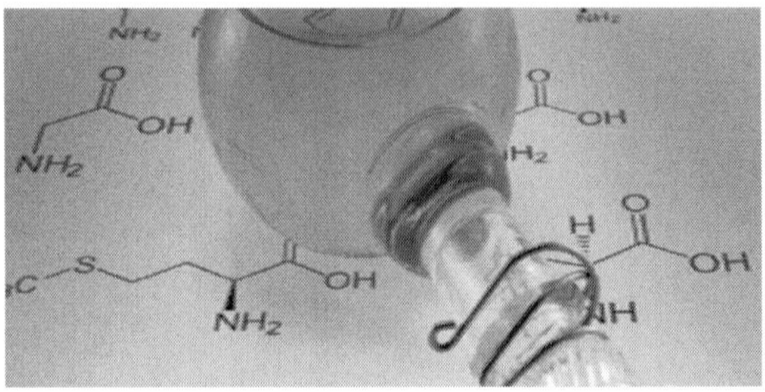

Figure 1.4. Water Chemistry.

Source: AquaNerd.

The ice can float in water has led to amazing use in the oceans, natural lakes, ponds and other areas where there is aquatic life or find their habitat, the ice acts as an insulating agent protecting the aquatic life below. If ice would be denser than water, it would mean that the ice would form on a pond and sink thereby exposing water to the cold temperature in the atmosphere and thus the aquatic species would rarely exist due to the harsh cold temperatures that they would be exposed to. This is because the pond, lake, ocean and other water bodies would freeze for long periods of time threatening the lives of all creatures present under the water. The substance is invaluable for transportation and recreation activities in that it is mostly found under liquid states under normal conditions. This also makes the substance suitable as a habitat for a myriad of plants and other animals.

Given that water can be easily transferred into the gaseous state, thus makes it possible to be transported to the atmosphere from the water bodies on earth. The transported water in gaseous form goes to the clouds and other inland areas where after condensation it pours down as rain. The rain in this form nourishes plants and other animal lives that are found on earth. Areas receiving higher rain have higher water bodies. This is because after the waters are struck by sunlight, they immediately change into gaseous form and evaporated into the clouds. Later the water in the clouds would go through the condensation processes and bring froth water in form of rain. These areas would be also termed as highlands due to the associated waters available both as rain and the water bodies present. The process is referred as hydrologic cycle, which would represent the process of transferring water between the land, ocean and or water bodies and the atmosphere. Religiously and philosophically, water has played a big role in human history's, in Greek philosophy water is regarded as the sole fundamental building block matter. In that, the matter encompasses everything including all forms of gods and men, atmosphere, mountains and hills, the earth's space, birds, grass and any living organism including the worms, flies and insects. All of them survive under different forms of water. All the religious aspects meditate on water.

Factually, about more than two hundred years ago water was considered and is still considered as one of the four fundamentals for existence. Accompanying water is the earth, air and fire or energy. The experiments done in the 18th century showed that the water is a combination of hydrogen and oxygen elements. Great research is ongoing on inventing a device that would harvest desert air and convert it into drinkable water. This would require a technology that would combine the hydrogen and oxygen elements available in the desert areas. On the earth's surface majority of the water is found in the oceans, this constitutes about 97% with only about 3%found in the bipolar ice caps and glaciers among them lakes, rivers and the underground waters contain freshwater, (Singh et al., 2021). Everyday world's population is increasing and in turn demand for freshwater is also increasing. Thus water conservation, recycling and purification are becoming the order of the day. The importance of conserving water is growing, as the available water is limited in supply. Purityof water required for human consumption purposes is way beyond the purity required for industrial purposes. Thus, recycled water is mainly used for industrial purposes. This is because water can carry other bacteria's that can cause human diseases in case of ingestion along with contaminated water. Certain diseases are waterborne for example cholera. This indicates that not all water is safe for human consumption. In addition to other sensitive areas is the case of high-pressure boilers where the water in use should at least be 99.9% pure for their effective use purpose. Waters found in the sea and oceans must be desalinated to be used for other uses because they contain large amounts of dissolved salts. Their use is limited in their remained states. They are mostly used for recreational purposes, as the water is very saline to be used for industrial and other human activities.

The structure of water is such that two hydrogen atoms are linked with a single chemical bond to an oxygen atom to constitute it. It has been reported from various scientific studies that most hydrogen atoms have a nucleus that consists of the proton, as the sole constitute. However, there are about two isotopic forms; deuterium and tritium in that the atomic nuclei normally contain one to two neutrons respectively are to some extent found in a small degree in water form. In chemical research Deuterium oxide (D_2O) is used as it is heavy water for better results. Deuterium oxide also known as heavy water is used as a neutron moderator in nuclear reactors. The single oxygen atom found at the outer shell of hydrogen to constitute water has six electrons. The electrons are used to hold a total of eight hydrogen electrons. Thereby, the two hydrogen atoms are bound to the oxygen atoms, the outer space is filled by the oxygen atom to give rise to water.

An experiment was performed in the past that showed that fresh water and seawater have different freezing points. The melting point of freshwater is about 0°C (32°F) while the boiling point is at about 100°C (212°F). The rates are much higher than would be experienced in the cases of other analogous compounds under the planet such as ammonia. This property is mainly attributable to the electronic structure that encompasses the water molecule. An additional fact about water is that the molecule is not linear but appears nonlinear or bent to some extent in a particular special manner. Moreover, two hydrogen atoms are bound to the oxygen atom at an angle of about 104.5^0. The oxygen-hydrogen distance is about 95.7 picometres (9.57×10^{-11} meters, or 3.77×10^{-9} inches) (Sroufe, 2018). Since the oxygen atom has larger electronegativity than the partnered hydrogen atoms, the bonds formed by the hydrogen and oxygen in water molecules are polar. The oxygen atom bears a negative partial charge, while the hydrogen atom bears a positive partial charge. The hydrogen bond is weak linkage that are formed by hydrogen atoms in the water molecules because they are attracted to areas or regions with high electron densities. Thereby, the hydrogen atoms located in the water molecule would be attracted to a nonbonding electron oxygen atom pairs on the adjacent water molecule. It is believed that water in its liquid state consists of aggregates of molecules in form of ware that keep reforming themselves. Water is thus found to have high velocity and large surface tension. On non-polar surfaces, the water droplets or beads form because the water molecules are adhering to each other than specific water molecules adhering to the surface.

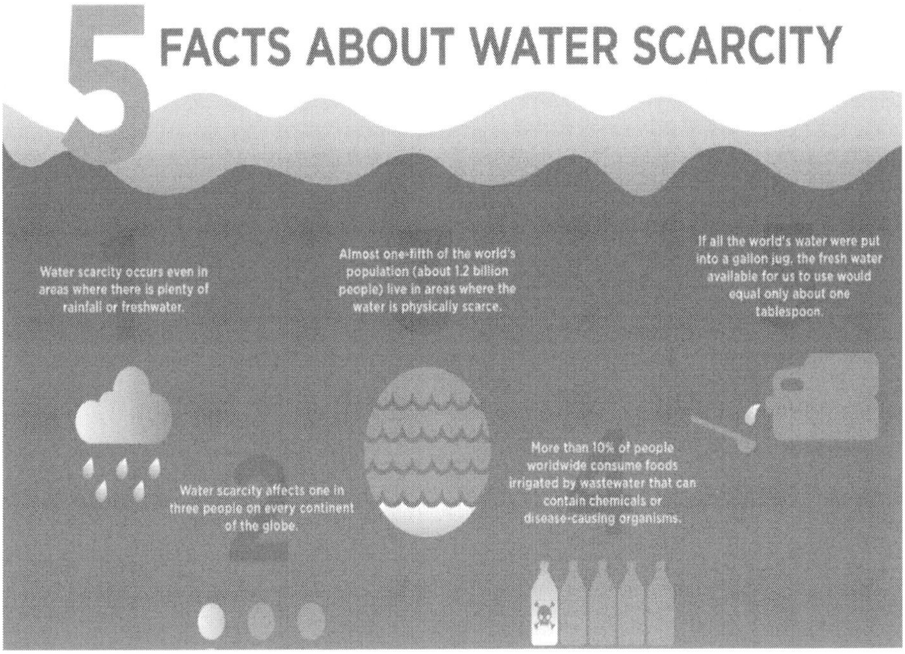

Figure 1.5. Facts on the world water scarcity.

Source: MECO.

Water is a renewable resource. The renewability however is limited by the demand that has been increasing worldwide over centuries. The population is a major driver that determines the availability of water that is in supply. In between the 1970s and 1990s, the per capita water supply is reported to declined by about a third. It is projected that water supply will continue to decline with increase in demographic growth. In the global view, water accessibility and availability have strained many continents in particular the countries in Asia because it accounts for about 60% of the world's population. The available supporting water sources in the world to this continent are about 36% (Prinz & Singh, 2000). Followed by Europe, which has about 13% of the earth's population with only about 8% of the world's water sources supplying water to its countries. Africa however has 13% of the world's population and accompanied by 11% of the words water recourses. On the contrary, the northern and Central American continents have 8% of the earth's population but have about 15% of the world's bodies supplying water to them. In addition, Oceania has almost 1% of the world's population but 5% of the world's water bodies. Thus, the geographical location of water is distributed unevenly leaving a majority of people with serious water scarcity challenges. The places having water instatic or flowing, permanent or temporary in marsh areas is termed as wetland.

1.5. USES OF WATER

Water is the most important natural resource that is required by all living beings. Both animals and plants need water for their survival. The quality of water and also the availability of water will determine the type of animals, plants or organisms reside in the particular area. Water will determine where people live. Although in some instances water has not always been available when needed and it may be available but not of good quality. Water as a resource to some people can be seen to be infinite but it should be regarded as finite as it will not always be available when needed. Water has its own limits and boundaries based on its availability and its suitability to be used.

Figure 1.6. Water, which is vital in our lives. Is considered finite and will always be available when needed.

Source: Cape Business News.

The balance between the demand and supply is quite delicate. Water is available for use is always determine and dictate to what extent development will occur. An area can never develop if the water demand outweighs the supply of water. Water can be needed for different purposes and uses and some are off-stream uses, i.e., water in this instance is withdrawn from a source and then there is the instream uses, i.e., water remains in place. In the past, countries such as the United States of America focused more on expanding the supply of water and also manipulated the supply of water in the country to meet the needs of the users. Many dams were built especially during the early 20th century to increase the supply of water to a given place at the required time (Partha et al., 2019).

As more researchers are advising the age and era of building dams has passed and what is needed now is to manage the already established infrastructure. The dams that are already built should be effectively managed so that there will be a sustainable supply of this precious commodity. How water is used from day to day can help to understand the effect of population growth on to demand and supply of water. This can help authorities come up with regulations, conservation policies and water management policies.

Although when most people think of water that is being used, they think of freshwater only thoughsaline water can also be used for different purposes. Some industries use saline water, industries such as thermoelectric, mining and few industries especially involving cooling generators.

There are different categories of water use include commercial, domestic, industrial, livestock, mining, public supply, thermoelectric to make power etc. All these different categories require water for their day to day running.

Commercial Water Use

Commercial operations are those that do not necessarily have an end product but they provide a service to people. Such commercial ventures include hospitals, restaurants and even schools.

Restaurants use a lot of water to operate daily. They require water to wash the establishment because it is a place to feed it has to be clean. The water is also used to cook food in the restaurant. Utensils that have been used need to be cleaned and rinsed to be used the next day or by the next customer that means water.

Vehicles need to be cleaned especially the vehicles that used by the public. These vehicles have to be cleaned regularly to maintain hygiene inside the vehicles. Motels and hotels also use water for their day-to-day activities, i.e., they need water for the customers to use such as to shower. After an individual has used a hotel and left or in the morning as they have woken up the sheets they used the previous night are taken and clean ones are placed. The dirty clothes are taken to the laundry to be cleaned. Water is used to clean dirty clothes (Owusu et al., 2016).

When there are wildfires and even fires that are extreme and individuals cannot put it out then they call the fire extinguishers or the fire brigade or in some countries the fire department. This extension of the local government has the responsibility of putting out fires. They use vehicles fitted with water tanks and use hoses to spray water onto the burning building or area. In some towns, there are water points where the fire department fits their hoses and use them to extinguish the fires. Water at this point is very effective as compared to other liquids or materials.

Figure 1.7. Fire extinguishers using water to put out a fire.

Source: The Times of Israel.

Recreational fields, green parks and golf courses are areas that essentially maintained green all year round. They need to be well maintained hence they have to be sprinkled to retain their aesthetic value and individuals who want to relax in these recreational gardens find it aesthetic and relaxing. Some countries have started making pitched and fields especially for outdoor games such as football, rugby, hockey some tennis courts and cricket fields that need to be watered to maintain their standards and also for the grasses not to dry and also to make the pitches wet so as avoid injury to the players.

Public areas such as streets roads and some parks have to be cleaned to maintain the. Most towns and cities have departments responsible for cleaning the town and/or city.

The town/city would be very stingy with mud and dirtif not clean on regular basis which requires a large amount of water. Public utilities such as toilets also require a lot of water for their cleaning (Olmstead, 2014). Some cities and town such as Las Vegas have many fountains, which help in the beautification of the city. The fountains require jets of water under high pressure and some are released in patterns, others are lined with led lights which come to life mostly at night and these have become attractions to the city. There are water parks such as swimming pools, splash pads, water slides and lazy rivers, which are specifically constructed for the amusement and entertainment of visitors. A lot of water is used in these parks but in most instances they reuse the water and purify them daily to maintain hygiene.

Industrial Water Use

Industries require a lot of water for their day-to-day running. Most of the time machines are coled down using water. Industries use water in various processes such as cleaning, transportation, dilution, cooling and processing. Industries that consume a lot of water include; food, paper, refined petroleum and metal industries.

Semiconductors are manufactured using water. Semiconductors that are very vital in computers, mobile phones and vehicles. Ultrapure water is used in semiconductor manufacturing to avoid interference by impurities present in normal water. Impurities such as silicon are removed from the water. This process is very expensive as 8 inches wafer which is a foundation for almost 100 chips requires at least 2000 gallons of ultrapure water (Olsson & Andersson, 2006).

Water is also used for cooling purposes in different stages of processing. The water is used to remove heat from different processes and different equipment's. Heat is therefore transferred from one medium to another. According to the SUEZ water technologies handbook, the driver of heat transfer is the difference between the two media that are in contact. Water has different properties and the behavior of the different compounds present within the other media. These compounds can be affected by temperature. This then brings about the tendency that a system can corrode, increase or reduce the scale or even in some instances support the activity of microorganisms and all this can be affected by the temperature of the water that is used.

The oil and gas industries consume a lot of water. They use water in different areas. Water is used during the process of drilling where water is used to cool the drill, to remove the drilling mud and rock debris on the drill, in hydraulic fracturing processes in that the water is mixed with other chemicals to improve its ability to create the fractures in the rock and also with sand to hold the fractures open and this will allow the oil and/or gas to flow into the well. In the process of oil sands mining, the water is heated and hot wateris used to separate bitumen from the sand and the clay.

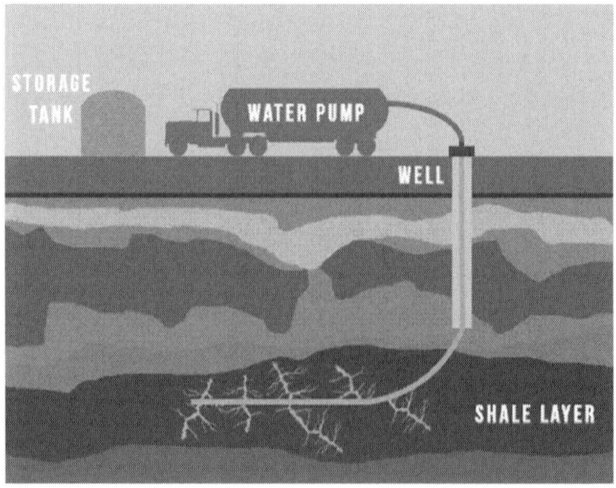

Figure 1.8. Hydraulic fracturing process. Water is pumped in between rocks to make them loose.

Source: wikihow.

Water is also used in paper and pulp mills. Although there has been digitalization and most people are moving from paper to digital technology there are still people proffering paper. In all the three stages in paper and pulp industries, 85% of water is used. Processes such as cooking, bleaching and washing all use water. A high amount of water is used and most of it is in processing and which leads to the production of a lot of wastewater (Nazemi & Wheater, 2015).

For some industries such as medical products manufacture, electronic goods makers and some food industries use deionized water as an essential ingredient. The deionized simply means that all ions in the water have been removed.

Irrigation Use of Water

Irrigation is the artificial supply of water to farms, orchads, and pasture or for horticultural crops. This may involve application of chemical, plant cooling, harvesting, freeze protection and leaching of salts from the top soil to the root zones. Irrigation is very vital as it helps in the production of food throughout the year irrespective of the season.

Figure 1.9. Chemical application of herbicides and/or pesticides. Water is used in the mixing of chemicals.

Source: Britannica.

During the cold season in areas such as Michigan, the plants can freeze hence the use of irrigation water. Water is sprinkled using a spray on the plants to prevent freezing of the plant. The idea is to use the energy that the water releases when it wants to freeze to change from solid to liquid to keep the ice right at the freezing point. This is to keep the ice wet then the temperature will remain at about 32° Fahrenheit.

Fertilizers and salts are dissolve in water before applying to the plants to improve the productivity and health of the plant. The solution of fertilizers and salts prepared in water can moves down the soil easily to reach the roots where the roots can now absorb the salts and other applied nutrients.

In the chemical application, the water and chemicals are mixed inside large tanks. The chemicals are mostly herbicides or pesticides used to kill weeds and pests that generally affect plants productivity. The chemicals after being mixed with water are then delivered to the plants via drip irrigation or through sprinklers (Muller, 2015).

The cooling of plants using irrigating water is performed during the hot months. Plants absorb water from the soil hence the water is delivered with pipes to the plant and as the plant absorbs the water from the soil it is cooled and the water from the plant that is hot is released through a process referred as evapotranspiration. The movement of water within the plant helps in cooling and thereby preventing the plant from wilting.

Irrigation is mostly carried out countries having dry enviornment. The countries want food security. Water from big rivers and even from boreholes are used to water the plants. In Kenya, plant crops such as rice is grown using water irrigation system. Rivers such as river Nyando in Kisumu and Nandi counties in Kenya supplies water which is used in basin irrigation so as rice can be planted. Large projects such as the galana-kulalu project where the Kenyan government is projecting the planting and use of a million acres. The plan will be irrigated with water obtained from the river Tana as the area is generally dry most time of the year (Afshar et al., 2015).

Some nonagricultural areas are also irrigated, i.e., cemeteries to maintain the green color of grass and to some cultures maintain it to show respect to the dead, turf farms, nurseries.

Domestic Water Use

Water is used in each home, as it is an everyday need. Water at homes is very crucial and anyone has to touch water or come in contact with water at home. Water at homes is used for many purposes such as drinking, food preparation, bathing, washing clothes, washing dishes, flushing toilets, watering lawns and watering the kitchen garden.

Water is used both indoor and outdoor. In homes, water is used for drinking is mostly bought. Water is supplied in containers ranging from twenty liters to the maximum that the individual needs. Drinking water is mostly placed on machines referred to as water dispensers which dispense water when a button is pressed and a cup placed below the tap. Most individuals do not drink water directly from the kitchen tap because they view it as being dirty and not pure.

Water is used for food preparation. Human beings mostly consume cooked food that is prepared in water. Water is very essential for food that needs boiling or simmering. The water is mixed with other ingredients to moisturize the other ingredients. Vegetables and fruits that are brought home are generally consumed after cleaning. Individuals are advised to wash the vegetables and fruits using running water to remove the germs and bacteria and other microorganisms that might bring diseases such as cholera.

Water is used for bathing. Human beings are advised to take a bath daily. Individuals take showers while inside their homes and special rooms are set aside and enclosed to hold the shower and those who are well off financially have a bathtub. Water is supplied to the bathroom

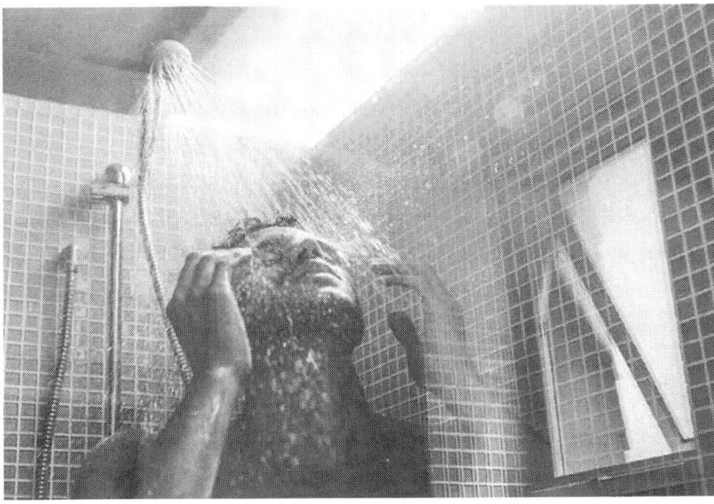

Figure 1.10. An individual taking a shower. Water is used and it comes from a showerhead at the top.

Source: Healthline.

via pipes and leaves by a shower. An individual takes shower with water from the pipes after scrubbing themselves with soap. In a bathtub, an individual opens the tap and water is filled up to three-quarter of bathtub, adds soap solution to form foam. An individual enters and soaks themselves into the bathtub. In countries that are still developing most individuals, take their baths along riverbanks where men and women have their designated spots to have a bath.

Dirty clothes are usually washed using water. After individuals have worn their clothes and attended to their daily activities, be it at the office, in a company or even a construction site, all change their clothes for cleaning the previous set of clothes (Ludwig et al., 2014). Especially for individuals who work in areas where they come in direct contact with dirt, e.g., Mechanics, construction workers etc. who come in contact with soil, oils and any other material needed their clothes to be washed. Water is used to clean clothes in combination with bleach and soap. Some individuals have washing machines at home, while others go to laundromats to clean their clothes. For individuals who are not well off financially clean their clothes using basins with water inside. In developing countries, people use water from the rivers to clean their clothes.

Utensils need to be clean every time after they are used. Individuals can use washing machines to wash their utensils. In areas where individuals have neither washing machines nor sinks to clean their utensils, they use their basins to clean their utensils.

Individuals have their basic amenities inside their houses and the toilet and shower are placed in the same rooms. After individuals visit the toilet either to urinate or to defecate they have to flush the toilet so that the waste can be taken to the septic tanks or into the sewer system. Water is used to clean the walls in the toilet so as not to attract flies into the house. In developing countries, individuals mostly use pit latrines. Although they do not flush the toilets and they have to clean the floors and walls regularly to ensure hygiene and cleanliness. The lawns and flowers are watered to maintain them. Individuals like their lawns and flower arrangements to remain green and beautiful so they have to water them. The lawns are usually watered in the mornings and evenings to ensure they absorb the water and remain healthy. Some families have their kitchen gardens in the backyard where they plant some vegetables and fruits. They usually water the plants to ensure they remain healthy and do not dry up.

Some house have their own pools in the backyard. The swimming pools are designed for relaxation and leisure for the family. The pools can sometimes be heated according to the needs of the individual. The pools are filled with water. Some more wealthy families have gazebos in the backyard where they relax mostly after a hard day work.

Farm Water Use

Animals that are kept in farms for monetary value also need water. Animals need water for day-to-day activities. Animals that are kept in the farms such as goats, sheep, pigs, rabbits etc. require water so that their metabolic activities can be carried out without any problems. Animals such as cows that produce milk need water because milk is liquid. On the farm, water is needed when livestock is being sprayed with solutions that kill pests such as ticks. Water is mixed together with acaricides and sprayed on livestock skin

Construction in the farm is always constant and many construction sites need water during mixing concrete water has to be added for the mixture to stick. After animals have used the various barns and crushes, they have to be cleaned. Water is used for cleaning farm structures to prevent the accumulation of dirt. Dairy parlor kept clean by washing it regularly to avoid spoilage of milk by the dirt and serious loss to the farmer..

Figure 1.11. Cows drinking water in a farm.

Source: Dairy Australia.

Water is used for cleaning utensils and containers used in the farm. Some containers such as milking buckets, milk churns, feed containers etc. have to be cleaned to ensure they do not transmit diseases to the animals on the farm. The water is also used in the post-harvest stage, to clean the produce such as carrots, potatoes etc. produce to attract customer and promote the sell (Lopez et al., 2009). Water is also used for cleaning tools that are used on the farm. Tools are washed to maintain their working condition.

Mining Water Use

Water is used in mining process during the extraction of minerals. Water is used during the dampening of roads to reduce the rising dust especially by lories carrying mined materials. Water is used for the transportation of ore and waste in slurries and suspensions. After mining has taken place water is mixed with ore and transported outside of the mine and water is filtered out or the suspension is left to settle and water is poured out. Water is also used together with other chemicals to separate the minerals in different chemical processes. Water is sometimes mixed with the mined materials and centrifuged to separate the materials into different layers. Water is also used in the cooling of machines used in the mining process. Water is used in the washing equipments used during the mining process.

Hydroelectric Power

Water is used in the production of hydroelectric power. This is a form of renewable energy that harnesses the power of water in motion. The name hydro- means water. This is electricity that has been generated from the flow of water. This has been used for several millennia as there is evidence that shows that the Greek used a mill that used water to turn the mill and in turn was used to grind wheat into flour (Leahy & Anderson, 2008). Most of the hydroelectric power plants have a large reservoir of water, held by a wall hence forming a dam. There is a gate to control the amount of water leaving the dam and an outlet where released water flows. Water that is in the dam flows as it has potential energy and the potential energy is then converted to kinetic energy. The water is used to turn the blades of a turbine hence generating electricity. The electricity is then supplied to the homes for use. The largest hydroelectric power plant in the world is the three gorges dam in china.

Thermoelectric Power

Here electricity is generated from heat. The source of the heat could be fossil fuels or from the earth's surface, i.e., geothermal sometimes nuclear fusion. Electricity is formed when turbines are turned by using steam. The steam moves at a very high speed hence moving the turbines. After the steam has been used to turn the turbines, it is condensed and it turns back to liquid form. It is then reused by rerouting it back to the boiler and it turned back to steam and the cycle is repeated.

Transport System

Water has been used as a transport system. For example, most of the nonperishable goods are transported on water, i.e., the seas and oceans. Some countries have also constructed waterways such as canals, e.g., the Suez Canal, tunnels and aqueducts. Most countries in Europe have inland waterways also known as canals where individuals use water for transportation.

Figure 1.12. Waterways or canals are used as transport ways in some countries.

Source: What is Danny doing?

Saline water is a liquid that contains high amounts of salts and other minerals. Although most of the water that is used is fresh, from springs and mountains while, some of the water that is used is saline. In the United States close to 13% of the water that is consumed or used is saline. Although saline water can only be used for certain purposes. Saline water cannot be consumed by animal or human being as it can harm their health. Saline water cannot be used for irrigation as it will wither the plant and eventually cause drying. Saline water is mostly used in thermoelectric plant cooling. It is used to cool machines that are used in thermoelectric plants. Saline water is also used in mining processes. Some individuals living along the ocean coasts use water from the ocean and this means they have to desalinate the water. The process is quite expensive and that the reason it is not commonly used (Larson & Lach, 2008).

Water use is increasing by the day. This is brought about by the fact that the population of human beings is increasing. The demand and supply of water are changing drastically as the demand for water is increasing and supply is constantly reducing by each day. According to research, the water use by the public and domestic will increase as years go by and increase in population, in the United States for example many sources of water are being stressed and this is because there are a lot of diversions on rivers, reservoirs and lakes and aquifers also have diversions on to meet the water need of homes, big cities, large farms, and industries that are ever-increasing. Environmentalists are also fighting for water to be conserved and leave the water from rivers and streams so that they can support the lives of fish and different aquatic organisms.

Although water is supplied to homes and various industries, all are always advised to harvest their own water and store it for future use rather than depend on water being supplied by the companies and local governments.

1.6. SOLUTIONS TO WATER CRISIS ON EARTH

Water is undergoing a great crisis due to the way of life human beings have adopted from the period of the industrial revolution. As the years go by, most of the industries are anticipating the transformation of the same meaning that the water will be at more risk than it is at the moment. Experts even claim that with the increase in the amounts of water pollution, there will be water shortages and the public will shift from a perception of looking at the water a commodity to exploit to a precious resource that needs to be taken care of. Jeff Erikson, a senior vice-chairperson at Sustainability, advised

Circle of Blue that the selections executives make to reply to freshwater shortage would penetrate nearly every factor in their enterprise operations. The numerous answers found out the complexity of handling water shortage such as population growth, city development, farm development and weather alternate are growing opposition for clean water and generating shortages. Here are the primary aspects in which specialists sense wanted solutions in.

1.6.1 Educate Individuals On How To Change Water Consumption Patterns and Lifestyle

In the process of ensuring that change occurs in any aspect, education always proves to be an important component. With education, individuals will have the ability to change their perception towards something and further promote motivate new ideas.

To cope up with the coming terror of water shortage around the world require that all hands be on deck on the same. It will require individuals utilize the supply chains that belong to major corporations. Some regions in the world are already facing water crises including the southwest US, Australia and India. One most critical thing is ensuring that individuals around the world are made sure to understand the problem of water shortages and its percussions if the same is not taken into account.

1.6.2. Need To Develop Energy Efficient Desalination Plants

Desalination has been a power-extensive method to curb the problem of water scarcity. Typically, the Middle East has capitalized on its big power reserves to construct desalination plants.

However, Saudi Arabia can be fostering a brand new form of desalination with its latest statement to apply solar-powered plants. Britain has taken an extraordinary method with small-scale centers for agriculture. Nevertheless, those improvements bring about every other wanted resource such as the capital for technological experimentation.

1.6.3. Invention of New Water Conservation Technologies

Water conservation is the next technology that will be able to save the world from the water crisis that is befalling people in the world. In most parts, aquifers levels and the rainwater have become unpredictable and thus the need to chip in innovations that can be able to save the world from the brink of falling. Individuals should try to cope with water scarcity, through the development of water conservation mechanisms, energy consumption. As the world strife to conserve water, they should also consider the energy type used in the process. Methods of energy that are used should be non-harmful to the environment. It means that while good is being done, it is vital to ensure that other aspects are not being jeopardized.

1.6.4. Improving Irrigation and Agricultural Practices

As seen earlier, some 70% of the world's fresh water is used for agriculture. Improving irrigation can play a vital role in narrowing the gap between the supply and demand of products.

In other instances, profligate irrigation practices supposed for an advanced generation has weakened the capacity of farmers to offer meals and fiber to a developing world. Examples consist of the Murray-Darling basin in Australia, Central Asia's Aral Sea, and the American Southwest (Kirschke & Newig, 2021). Although improved technologies have become attractive solution International water specialists, Peter Gleick notice that during a few instances, along with the rural structures in California, achievement tales can show up through enhancing what is already in place.

With the increase in the industrial revolution, the climatic conditions are changing for the worst and mostly drought, famine and in some instances are seen to befall the world.

1.6.5. Recycling of Wastewater

Water recycling just like plastic recycling is important to the world. It helps ensure that there is little water loss and need individuals to be very cautious in while using water. Water treatments are now being undertaken in the world and as such, individuals should be careful of what they dump in the waters. Taking for instance plastic bags.

These are some of the worst components that may prove to be a menace in the process of recycling water. The areas of recycling, should not have any form of blockages to ensure that water flows smoothly. Individuals need to be explained about the things that need to be done to ensure that the process runs smoothly. Some countries in the world have begun the process of recycling water to be able to cut water imports which have proven expensive to their economy. But it cannot be avoided in case of water shortage as water is an essential component of human life.

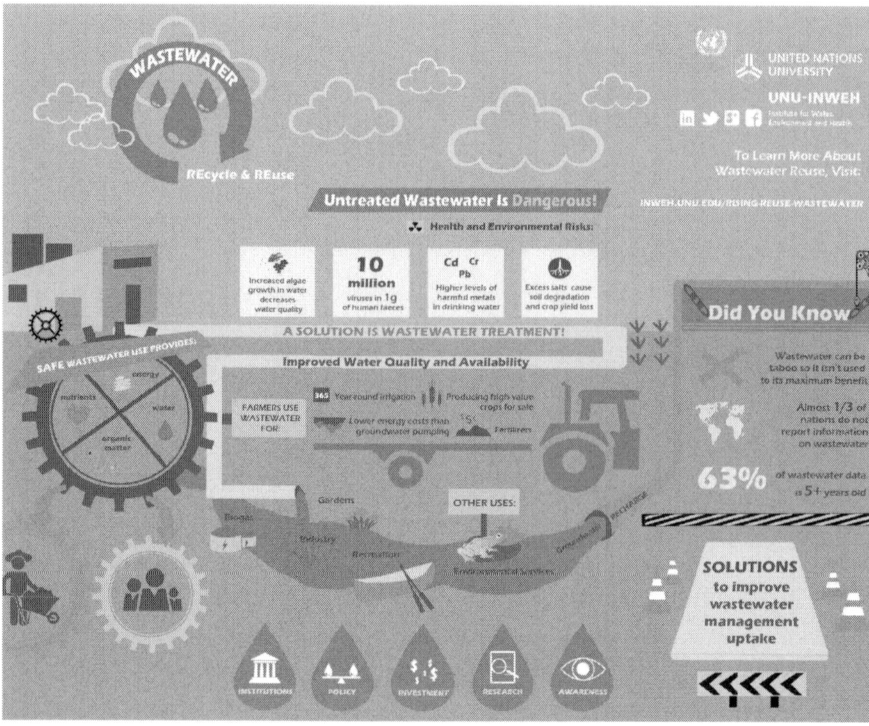

Figure 1.13. Water recycling and reuse. Through the process of recycling and reusing water helps in the process of overcoming the global water crisis.

Source: Genesis Water Technologies 1.6.

1.6.6. Development of Better Policies, Regulations and Enacting Them

As water shortage complicates meals safety and pollution, governments want to redefine their role. The U.S. authorities are thinking about increasing the Clean Water Act to make certain extra protections. In Russia, meanwhile, Prime Minister Vladimir Putin authorized waste discharges in Lake Baikal, one of the world's biggest bodies of freshwater. Regardless of what direction elected leaders take–the Circle of Blue/GlobeScan WaterViews survey shows they're thinking about a couple of approaches. The survey additionally discovered that maximum human beings say it's far as much as the authorities to make certain groups to get easy water. Leaders are the determinants of how a country is being run. As such, it is important that as leaders, the choice made for the country works for the good of the citizens.

Water is very important and to prevent cases of diseases and hunger, available water should be handled with much care and should be protected from substances that may harm the users of the same. Individuals need to be cautious about some of the policies made to ensure the safety of the water consumed by individuals around the world.

1.6.7. Looking To Community-Based Governance and Partnerships

Communities tend to work together well in the case where they need to achieve certain objectives. Taking for instance the case of protecting a particular water body. In this case, all the members in the particular community know the importance of protecting their resources and they will take an initiative to protect them. With community organizations, the voice of individuals who are experienced in do and don't with regards to water conservation will be heard. Taking for instance with the case of climate change where individuals are willing to bring influence to the society of what needs to be done to approach such a menace. The organizations tend to mobilize individuals and assist the individuals with knowledge and experience of the best methods to approach a given crisis. When there is effective governance at the grass-root levels, there is a capability of effective policy changes at the national level.

Figure 1.14. Community-based organizations in action. These organizations are quite useful in mobilizing individuals within the community and also ensure that influence is provided to the individuals with knowledge of what needs to be done.

Source: OpenGlobalRights.

1.6.8. Management of the Ecosystem Holistically

Simply put, holistic control applies to a practical, common-experience method to overseeing plant assets that take into consideration economic, cultural, and ecological goals. In essence, the complete is more than the sum of its parts, and every aspect is associated with it and its impacts on the others. Good examples of holistic control are groups that function sewage remedy plants even as pursuing partnerships with easy power manufacturers to apply wastewater to fertilize algae and different biofuel crops. The crops, in turn, absorb vitamins and purify wastewater, substantially lowering pumping and remedy costs. In a way, it can be seen that the ecosystem has a way of remedying itself and further ensuring that it thrives. Let us just say that without human life, the ecosystem would have thrived or rather with human presence and without industrial revolution, the thriving would be evident (Krampe, 2017). The only thing left to do is to try helping the ecosystem restore itself given the fact that humans are the major contributors to disruptions in the world.

1.6.9. Improvement of Water Catchment and Harvesting

In most cases, water catchments are normally useful in areas where water is considered as a scarce resource. Taking for instance Pakistan and India, which are some of the areas that are facing difficulties due to climatic changes. As such, the two countries have been able to construct dams and other water harvesting methods to help them harvest rainwater. With the presence of water catchment areas, individuals can be able to conduct agricultural activities even in instances when rainwater is not available.

1.6.10. Improving the Distribution Infrastructure

Poor infrastructure is devastating to human beings, especially when dealing with their health and the economy at large. Good infrastructure reduces the wastage of resources, reduces transaction costs, the high satisfaction of life, and prevent water-borne illnesses to unfold amongst susceptible populations, mainly children and the old members of the society. The trouble is not always constrained to the growing world (Kim et al., 2015). Pipes burst on an ordinary foundation taking for instance in the U.S., prompts boil alerts. Sewage remedy structures often overflow and malfunction, inflicting seaside closures which then lead to bad odor and risks of diseases.

1.6.11. Building of International Frameworks and Institutional Cooperation

The process of building accords for the main purpose of ensuring the aural resources are protected always proved to be a hard process. In 2009, a conference that was held in Copenhagen by the United Nations involving climate change proved this aspect was true. International frameworks are important, as they are bodies that can help ensure that individuals can follow the rules regarding the use and utilization of resources. The problem that accrued did not just affect the water crisis but also other climatic factors. Furthermore, dealing with issues related to water bodies that are shared between or among states also has its challenges given the fact that individuals always find it hard to reach an agreement. However, policymakers and advocates are always cutting effort to ensure that the process will one day become a success. UN is considered as one of the Humanitarian-oriented treaties and has proven that the achievement of millennial development goals including drinking water is achievable.

1.6.12. Address Pollution

Pollution is a very common issue in the world. Water pollution is becoming problematic, as it has led to increased diseases in the world and an insufficient amount of water for the people in a community. Most industries depose their waste to the water bodies, mainly rivers that in most cases are used by individuals downstream. This happens especially in the underdeveloped parts of the world where most of the people do not have access to water in their household and have to go to the river to fetch water. Measuring and tracking water is vital to human fitness and biodiversity. This huge problem rears its head in lots of paperwork and may be addressed in only as many ways, whether or not it's considered as David de Rothschild's eco-journey in a plastic delivery or Joe Berlinger's documentary on oil contaminating the Ecuadorian Amazon. While securing the high satisfaction of consuming water and on the neighboring level, it is vital to construct worldwide bridges to solutions.

1.6.13. Controlling Population Growth

In the world, the population is always growing and is risking many important resources. One factor is that; the supply of resources is becoming limited and the demand is rising by the minute. In the current world, approximately a billion people do not have access to clean water. As such, nations need to come up with measures to control population growth. If population growth is mitigated in time, resources in high demand will reduce which will bring about the age of terror and war where individuals will be fighting for their right to access water resource.

1.6.14. Water Developments in Developing Countries

Developing countries are in the process of getting towards a development age. The problem of climatic change and scarcity of water are the aspects that have brought about dramatic consequences. The scientists have proposed solution to this problem is increasing the transfer of conservation technologies especially for areas that are affected by dry weather. Imlementing such technologies are proving very difficult due to weak economy of developing nations are known to have (Jackson, 2006).

There is no doubt that, water is important to human life existence and has proved to be very useful in many circumstances. However, it is facing a major crisis and the individuals in the world need to come together to deal with such challenges.

CHAPTER 2

WATER RESOURCES

CONTENTS

2.1. INTRODUCTION

Water resources are the natural sources of water that are very much important. Only 3% of the water on earth is freshwater, with little more than two-thirds frozen in glaciers and polar ice caps. The bulk of the water on the planet is saltwater. The most prevalent ions in saltwater are chloride and sodium. Additional ions or elements found in the oceans are magnesium, sulphur, calcium, potassium, and a variety of others (Hirji & Ibrekk, 2001). These ions combine to form salts, which give oceans, seas, and some lakes their salty or saline flavour. Organisms in these ecosystems have adaptated that help them to live in the constant presence of water and salt. Humans cannot directly use salt water, as humans feel nauseous upon consumption of ocean water. Saltwater need to be desalinized before humans use it for drinking or crop irrigation, in which salts are removed. Desalinization is expensive and energy-consuming, but in particularly arid areas, it can provide individuals with much freshwater.

The remainder of frozen freshwater is largely contained as groundwater, with barely a trace detected above ground or in the air. Natural sources of freshwater include surface water, subsurface river flow, groundwater, and frozen water. Two artificial forms of freshwater are treated wastewater (reclaimed water) and desalinated seawater.

Water is used in agriculture, industry, housing, recreation, and the environment. Water scarcity, pollution, water conflict, and global warming are all impacting water sources. Culinary freshwater is a source of renewable energy but the availability of groundwater worldwide continues to decline, with the most noticeable degradation happening in Asia, South America and North America andthe balance between natural renewing and ecosystems is still not obvious. Water rights are known as the framework for the allocation of water resources to water users (when there is such a framework).

Among other natural resources, water resources have a distinct place. Water is world's most widespread substance: it is present everywhere and plays an important part both in human existence and the environment, but in varying quantities.

Freshwater is of the greatest significance, human life is inconceivable without it, since nothing else can replace it. People have always consumed freshwater and employed a full range of applications with the varied natural surface water systems. The impact of man on water resources has been negligible and local for many hundreds of years. The beautiful features of natural waters – their refurbishment during the water cycle and their ability to purify themselves – have ensured for a long time that freshwater is relative to its purity, quantity and quality.

The direct use of water resources as well as surface alterations in numerous catchments of rivers. This has been ubstantially increased during the 1950s in worldwide water withdrawal. This growth, on the other hand, was due to the revolution in science and technology that allowed for intensive production capacity development in all domains of the international economy. The yearly water withdrawal between 1951 to 1960 was four-fold higher than in the preceding decades (Ghosh, 2017). This has been due to the spectacular expansion of irrigated areas, the growing demand of water in industries, heat-engineering, and the extensive building of reservoirs on all continents. The hydrological round of rivers and lakes, impacting their water quality, their water potential as resources and world water budget, has changed drastically all across the world for the past 25–30 years. The scope of water resources, their spatial and temporal distribution, is not only governed by natural climate changes as before, but also by the economic activities of man.

Water supply in many places of the world has gotten so low and contaminated that it is already unable to meet its increasing needs. This has been the greatest obstacle to economic progress and population growth.

Severe freshwater issues exist in dry areas with limited water supply, high levels of use and very rapid demographic growth in Argentina. The very first United Nations World Water Conference was held in 1977. This seminar examined the low quality of fresh water in a restricted rainfall area of one-third of the world.

It was also stated that in the majority of the countries by the end of the century many of the same difficulties currently present in these areas will develop. The conference of 1977 has made a major contribution to developing and coordinating worldwide cooperation in water resources studies and evaluation. It also promoted national research, development and brought numerous issues of management, as well as further growth in hydrological research to the general public's notice, the governments, interest groups and decision-makers. Regional and worldwide water resource concerns are the attention of various international governmental and NGOs such as UNESCO, WMO, UNEP, FAO, IAHS, the IWAR, etc. who have all sponsored several scientific congresses and symposiums to address these challenges. In various countries a huge number of investigations have been released, maybe with the most precise and full water balance estimations for each landmass and the natural geo-morphological zones of the Earth authored in 1974 (Grizzetti et al., 2016).

Furthermore, the dependability of the information is not regarded as very good as it originates from diverse sources, possibly outdated values for different years or other long-term periods. Consequently, within the framework of the International Hydrological Program, UNESCO developed a new initiative for the standardization of the world's water resources and usages and preparing a World Water Resources monographs in the early 21st century. The St. Petersburg, Russia State Hydrology Institute was responsible for implementing the project between 1991 and 1996.

In the natural setting, water is the most widely used material. In three states, water is available: liquid, solid and undetectable fumes. It produces the seas and oceans, seas, lakes, rivers, and subsurface waters in the earth's uppermost layers. In a strong condition, it exists in polar and alpine regions as ice and snow cover. Some water in the air, such as vapour, water vapour, ice crystals, and biosphere are included. In the constitution of the various minerals in the earth's crust and core, vast volumes of water are connected.

To correctly evaluate the total storage of water on earth is a complex problem since water is so variable. It moves from liquid to solid or gaseous form and then again to liquid. It moves continuously. The amount of water present in the so-called hydrosphere is usually estimated. This is all the free water that is released into the air, on the ground and in the rock at a depth of 2000 meters, whether in liquid or solid, or gaseous conditions. The hydrosphere of the earth holds a massive volume of water—around 1386 million cubic km. Nevertheless, saltwater accounts for 97.5% and fresh water only 2.5% (Geissen et al., 2015).

In the Antarctic, the Arctic, and mountain ranges, the majority of all this rainwater (68.7%) is in the state of ice and perennial snow cover. Secondly, like fresh groundwater, 29.9% exist. In lakes, reservoirs, and water systems, just 0.26% of the overall quantity of frozen water in the world is most readily reachable for economic requirements and critically essential for ecosystems. It is the amount of water contained in water systems, aquifers and the atmosphere at the same time, averaged, over a lengthy period. The quantity of water preserved in the hydrosphere will shift with water interchange between oceans, lands, and atmosphere during brief durations such as a year, a few years, or even a few months. As seen in the graphic below, this transfer is known as the earth's water circulation or turnover and more officially the global hydrological cycle.

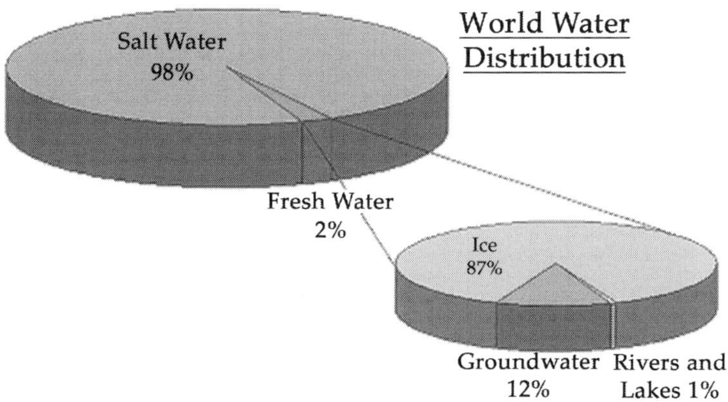

Figure 2.1. Distribution of water across the world.

Source: https://snr.unl.edu/data/water/groundwater/realtime/waterdistribution.aspx.

Solar heat drains water from the surface of the earth into the atmosphere. A constant stream of water vapour spreads out throughout the surface of the earth before it falls then as precipitation, on land, lakes, rivers, and oceans. The major source for the development of waterways on land, rivers, lakes, soil waters and glaciers, is rainfall or any kind of precipitation that falls into the land. The rest – as river flow – travels to the oceans where they evaporate: this is repeated time and again. Some of them infiltrate and energize the groundwaters.

A substantial part of the flow does not come into the ocean, as the areas without the natural rainwater pathways are evaporated in endorheic regions. And from the other extreme, some groundwater completely circumvents river systems and travels to or evaporates straight into the sea. The graphic shows quantitative indications of these many hydrological aspects of the global cycle. The earth's water cycle includes 577,000 km^3 of water each year. This moisture goes from the ocean (502,800 km^3) and the ground (74,200 km^3) (Gao et al., 2014).

The very same quantity of water vapour is dropped as precipitation in the atmosphere, 458,000 km^3 of ocean and 119,000 km^3 of land. The discrepancy in precipitation and land evaporation (119,000 to 74,200 = 34,800 square miles per year) is the total runoff from the rivers of the earth (42,700 square miles per year) plus the direct rushing of groundwater towards the ocean (2100 square km per year). These are the main sources of freshwater for the promotion of human life needs and business.

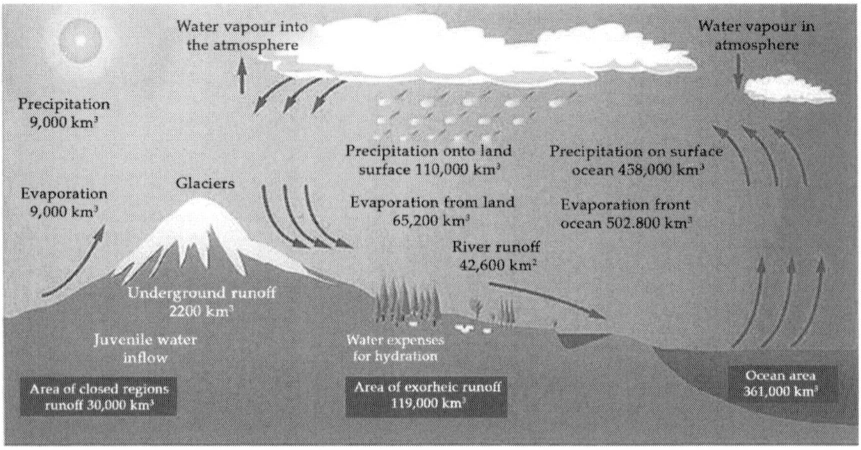

Figure 2.2. The world's water cycle.

Source: http://nwpi.krc.karelia.ru/e/climas/Runoff/Hcycle.htm.

Water on the river is extremely important for humanity and in the global hydrological cycle. This is because the constituents of the water turnover on the earth are complementary to each other's behaviour, both according to stock size and water dynamics. Even during the hydrological cycle, but fluctuate at varying rates, the diverse kinds of water in the hydrosphere are refilled totally. For example, it takes approximately 2500 years for the entire refill of the ocean water, around 10,000 years for permafrost and ice, around 1500 years for deep groundwater and mountain glaciers. Water is stored fully over 17 years in lakes and around 16 days in rivers (Fuhrmann, 2012).

Two approaches are widely employed in hydrology and water management based on water exchange features in a region to evaluate water resources: the static storage component and renewable water. These define the ideas of the types and characteristics of water sources. Static storage usually comprises of freshwater, for example, big lakes, groundwater, or glaciers, which is completely regenerated over many years or decades. The continuous use of this component inevitably leads to storage deterioration and unfavourable effects. It also upsets the natural balance established overages, which would take decades or hundreds of years to restore.

Renewable water resources are replenished on an annual basis as part of the earth's water cycle. They are usually rivers, calculated as volumes by unit time (a quantified percentage m %), km/year, etc.) and created in a particular location or from external sources, including the influx of groundwater to a river network. The annual renewable high aquifer groundwater not exhausted by river systems also contains this type of water resource. These quantities, nevertheless, on a global basis are not enormous with the amount of river runoff and are important primarily in particular places.

The river runoff is not only replenished quantitatively but also qualitatively during the turnover phase. Now if a man can cease polluting rivers, water would be able to recover to its natural cleanliness over time. As a result, river runoff is the most significant component of the hydrological cycle, as it represents renewable water supplies. It has a significant impact on the earth's surface ecology as well as human economic development (Figuères et al., 2003). The river runoff is the most broadly spread throughout the land surface and supplies the world's largest quantity of water use. In reality, the magnitude of river runoff is used to assess water availability and/or deficits for a certain location. As a result, the assessment of river runoff, its spatial-temporal patterns on a worldwide size, and a study of its usage for various economic requirements, both now and in the future, are the focus of the subsequent segment of this overview.

Two methodologies can be used to determine the quantitative properties of renewable water resources in a territory or river basin: meteorological statistics or river runoff measurements. When there are limited hydrological measurements but significant detailed and extensive meteorological data, the first strategy is often used. In reality, it's commonly accomplished by utilizing the simplest equation for a territory's long-term water balance. The contrast in streamflow from land, when summed over a long period of time, determines the magnitude or area of renewable water resources. Precipitation is determined using several estimation methods based on observational evidence and evaporation level.

This simple estimation methodology has been frequently utilized in many nations since the 1900s when the hydrological infrastructure was underdeveloped and climatic data was much more openly accessible. The methodology is still used relatively regularly; for example, it was used to estimate water resources for Latin American countries. Although straightforward, calculating sustainable water resources using meteorological data has several major drawbacks. It is not suitable for precise computations, particularly in nations and places with limited water supplies.

Firstly, the low precision means that this methodology does not apply to dry and semi-arid regions in which river runoff is very modest, as absolute numbers near any inaccuracy in evaporation and precipitation measurement.

Secondly, for individual years or, furthermore, for a particular season or month, it cannot be properly estimated water resources by this approach. Thirdly, this methodology is not suitable for the evaluation of water resources in international, nations, and territories, because a greater volume of the rivers flows beyond the area (Fang et al., 2013).

The second methodology has been used in this regard to calculate global water resources for geographies, states, and even continents in diverse physiographic situations. This is predicated on reports from the world hydrologic system, which is an auxiliary function of the meteorological information. Early in the 60's and 70's, this methodology was effectively used to prepare the World Water Balance and Water Resources of the earth as mentioned previously.

Figure 2.3. The World Water Day is an annual United Nations day for observation of freshwater, conducted on 22 March.

Source: https://www.worldwaterweek.org/event/8398-world-water-day-2020-water-and-climate-change.

An additional number of hydrological observations have already been gathered, and records hitherto unavailable have been obtained, which have not been covered with meteorological data for various regions of Africa, Asia, and Latin America. The application of this latter strategy is consequently significantly more justified. One of the primary prerequisites for choosing hydrological sites was the provision of a long-term monitoring series to satisfy a core characteristic that should allow water resources assessment across the globe to be based on an identical period in time. Regrettably, the selected sites in Africa, Asia, and Latin America, in particular in poorer countries, had very few entries (5 to 10 years). In fact, data in most sites was acquired only until 1980–88. The only study period from 1921 and 1985 has been selected however data are available for 1990–94 for numerous nations in Europe and North America (Dobriyal et al., 2012). With this quite strong 66-year design period, equivalent average water resources calculations may be produced in all parts of the globe. Their excessive values and long-term fluctuation might also be properly estimated. A great deal has been done to update the collection and repair observation holes such that continuous data are collected for the designated site for a long timeframe. This has been accomplished by well-skilled correlation algorithms and in many cases, weather data are used in order to generate more trustworthy results. Monthly precipitation and air temperatures of more than 7000 scientific sites utilizing the global weather network are collected.

Despite the relatively large global hydrological network, around 15–20% of the land area is not represented by routine survey data. These are the normally thinly populated arctic, desert, and tropical zones, and hydrological equations and modelling methods were utilized for annual runoff to determine runoffs from these areas. The water resources of specific countries and those natural-economic areas which do not match the borders of river basins must be evaluated in special ways. Hydrological models expressly created in this case at the Russian State Institute of Hydrology were utilized to create credible estimates of the main features of renewable water resources in specific regions. At present, 64,000 stations with measured river discharge are reported by the World Meteorological Organization. These are dispersed throughout the world extremely unevenly. The observation time varies between only a few months and over 180 years. More or less, river flux data is available from around 40,000 stations while the integrity of data from various countries, including those that are fragmented, is very diverse (Cook & Spray, 2012). There is also observational data available by which national or regional charts for yearly average river discharge were developed. It would be hard to use all this information to predict water resource trends globally so that hydrological facilities were selected based on the same conditions:

- The oldest active series of observations is available.
- The placement of sites on wide and medium rivers spread preferably equally throughout territories.
- The natural (or similar to the natural) river runoff cycle should be reflected in the observations.

As a result, the monthly and annual records from over 2,500 hydrological sites were used effectively to estimate the worldwide distribution of sustainable water resources on all continents. The number of locations in Asia is approximately 800, Europe approximately 600, North America 330, South and Africa from 250 to 50, and Oceania approximately 200 (Cabrera & Custodio, 2014).

2.2. CONTINENTAL, REGIONAL, AND COUNTRY DISTRIBUTION

The average value of world water resources from renewable sources is estimated to be 42.700 km^3 per year, yet is significantly changeable over time. As regards absolute values, Asia and South America are the biggest volumes of water resources (13,500 and 12,000 km/a). The smallest are usually those with Oceania for Europe and Australia (2900 and 2400 km3 per year). The water resources might vary in extent from 15 to 25% of their mean value for individual years. Pure figures do not completely reflect the availability of water in the continents, since they vary so greatly from one place to the next. Specific access to water, as illustrated in the diagram, is far more informative. The fast population expansion between 1970 and 1994 led to a decrease from a population of 12.9 to 7.6 thousand cubic meters per year per individual in the prospective water available for the earth population. In Africa (2.8 times), in Asia (2 times) and in South America, the population water supply has been cut to the maximum level (by 1.7 times). In the same era, water supplies for Europe's population declined only by 16% (Brack et al., 2015). (Note that the true water supply declined well above those numbers for many African, Asian, and South American countries when the account is taken of increases in water demand and this is covered subsequently.)

Research has indicated that changes in the overall river runoff over the entire earth on the continents have a cyclical pattern. Wet and dry years switch cycles and variations in time and intensity from the average values differ greatly. For example, times of low water levels such as 1940–44, 1965–68, and 1977–79, when values fell 1600–2900 km lower than average, maybe seen clearly in the overall flux of the world's rivers. In contrast, river discharge had increased significantly in 1926–28, 1949–53, and 1973–74.

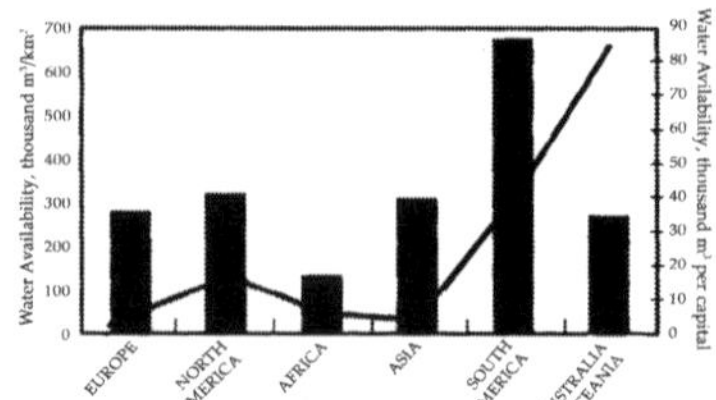

Figure 2.4. Table showing continental water availability.

Source: https://snia.mop.gob.cl/sad/PHI710.pdf.

Together with such strictly delineated cyclical changes in the total world-flow, it is also pretty evident that during the whole 65-year span there is no obvious pattern. It applies to all continents, especially if the statistics for Africa and South America during the past two decades, where riverbeds are increasing in South America and declining in Africa, are not taken into account. If the worldwide riparian flux over a long-term period does not grow significantly, doubts are to be raised regarding the broad idea that the world ocean has risen since the turn of the 20th century owing to increased continental mountain glaciers.

River runoff in practically all locations of the world is relatively imbalanced year-round. During flood periods, over 60–70% of the runoff is produced and consequently, renewables vary substantially over a year. It is now expected that much of the rush in will take place in Europe between April and July (46%), Asia from June to September (54%), Africa between September and December (44%), South America (45%) between April and July and Australia/Oceania (40%) during January and April. The rainy season for the whole land area is on average between May and August, with a total global fluctuation of roughly 45% of the total annually (Alamanos et al., 2018).

The disparity in the dispersion of river runoff during a year means that reservoirs of different kinds need to be regulated. For water supply, it is important what we term a stable volume of base flux, which changes little over a year and year after year. Without technological regulation, its usage is feasible. Its volume is roughly 37% or around 16,000 km^3 per year of the worldwide river runoff volume.

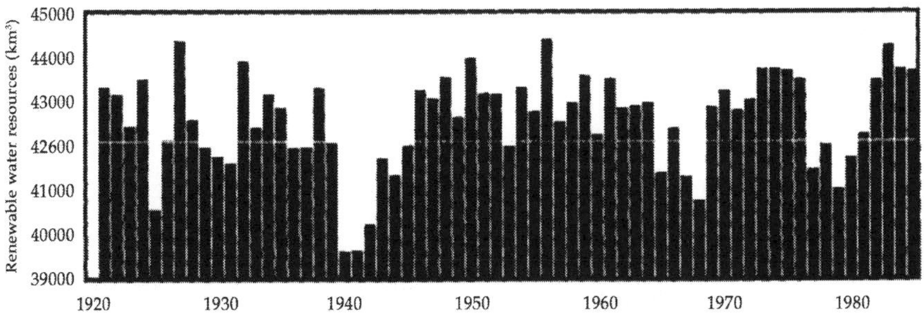

Figure 2.5. Renewable water resource levels in Km3 over the last 100 years.

Source: https://snia.mop.gob.cl/sad/PHI710.pdf.

Renewable water resources: temporal/geographical fluctuations. Let us now analyse in more depth the temporal and geographic changes of renewable water resources. For this reason, major regions with homogenous geomorphologic properties and a similar level of economic development can be picked from 3 to 8 in each continent, i.e., a total of about 26 unique regions. Regional boundaries have in most cases been administrative boundaries and so certain areas cover the whole territories of each country. This can be achieved by evaluating the number of water resources together with population trends and water use for which only national governments are disclosed. Russia, China, and the United States are exceptions, with some individual components in various natural and economic areas being combined. The individual areas range from 0.20 to 0.19 million km for Transcaucasia, from 12 to 13 million kilometres for Siberia, and for the Russian Far East, Canada, and Alaska: the majority of regions are 1 to 8 million kilometres.

In regions, year-to-year fluctuation can be very considerable and the average values for continents can be substantially higher. This is notably so in areas where the real values are tiny in arid and semi-arid regions. There, the readings maybe 1.5 to 2 times lower for each year than the long-term averages, while for wetlands this difference may be between 1.5 and 25%. Several long-term variances for Africa and Europe in renewable water resources. It has a cyclical character and dry spells can endure for 3 to 6 years in nearly all regions, with serious water supply issues in several areas (Butt & Bilal, 2011). This decline was particularly dangerous for the Sahel because water resources had fallen nearly half in the previous 15 years. Not just year-to-year but also seasonally and monthly fluctuation impact the potential of exploiting water resources for economic demands. A widespread riverside dispersion characterizes many countries with 60–80% of annual runoff during the wet season, which lasts from three to four months. In northern and southern Europe, for instance, 64% of the annual rainfalls occur during the three flood months. An estimate has also been given for renewable water resources in 60 individual countries

These include both established and developing nations and transitional economies, as well as those with the greatest and the smallest populations. Six major countries with the biggest renewable water projects include: Brazil, Russia, Canada, the USA, China, and India, which account for about 40% of annual river runoffs. Due to their huge geographical mass and different climatic conditions, the runoff dispersal over a year in these countries is significantly less.

2.2.1. River Basins and Runoff to the Oceans

The world's biggest river, the Amazon, contributes 16% of the annual worldwide river runoff; the 6 biggest rivers in the world include the Amazon, the Ganges with the Brahmaputra, the Congo, the Yangtze, and the Orinoco provide 27% of the world's water resources. Except for Australia, the flushes illustrated in the graphic are located in every part of the world. All three rivers make up 52% of the world's supplies of water (Bouwer, 2002). The freshwater influx into the world ocean could be assessed by generalizing river flow data from the world hydrologic circuit. This input is significant to investigate the equilibrium and dynamic mechanisms of freshwater.

However, freshwater inflow to the World Ocean cannot be identified simply based on the value of global river runoff for two reasons:

First, river watersheds mostly belong to the so-called endorheic (drain-less) runoff regions that are not connected to the World Ocean. The distribution of these regions is shown overleaf on a schematic world map. The total area of the endorheic regions is about 30 million km*, or some 20% of the total land area. However, only 2.3% (about 1000 km/year) of annual global river runoff is formed in these regions simply because most of these drains-less regions are occupied by deserts and semi-deserts with very low precipitation (Bouwer, 2000).

Secondly, the water resources quantities in river basins do not necessarily correlate with the river mouth rushing in the exorheic drainage areas that are directly linked to the World Ocean. This applies especially to the hot climate zones. In the mountainous areas with high precipitation, the basin's water resources are generated. Going towards the outlet of the rivers, the evaporation of the flood plains and lowlands is lost due to a lot of run-offs. All these River Basins can be found in Ganges and Indo, Asia, Niger, and Zambezi in Africa, Mississippi, and Colorado in North America.

Together with all the exorheic areas, some 1,100 km of runoff is wasted each year as it evaporates and does not approach the mouth of the river. The world's biggest, i.e., rivers with a long river volume averaging more than 100 km^3 (Bagatin et al., 2014). The entire flow of water in the world ocean will therefore be somewhat lower than the quantity of all the continents of the world's renewable water resources. The Atlantic Ocean, which has four of the six largest streams in the world, is the Amazon, Congo, Orinoco, and Parana, which account for around half of total river water flows in.

The smallest amount – some 43,000 kilometers per year – of river water is flowing into the Arctic Ocean; yet, river waters are extremely important to the ocean regime. This may be explained very clearly: in the Arctic Ocean 1.2% of the entire water supplies are stored in the World Ocean, but, simultaneously, 11% of the worldwide water supply is acceptable. It is highly crucial to consider not only the quantity of river water intake but also its dispersion to replicate the dynamical mechanisms in the oceans since the river runoff is entering the world oceanquite unevenly. In the equatorial area between 10" N and 105 on average, approximately 40% of total river ripples join the ocean.

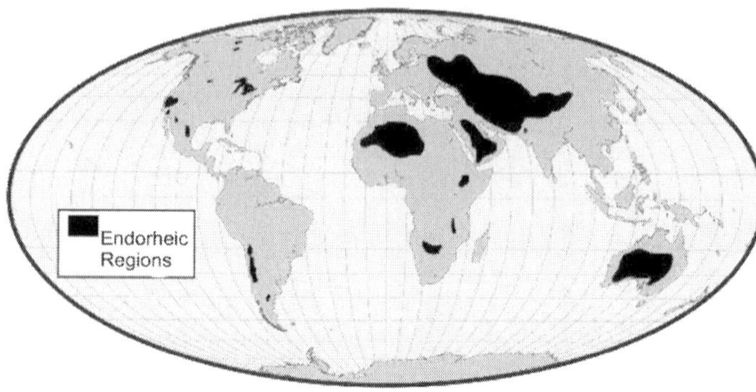

Figure 2.6. Global endorheic zones.

Source: https://link.springer.com/article/10.1007/s10040-020-02209-8/figures/1.

2.2.2. River Runoff and Groundwater

Total river discharge estimates the size of the renewable water resources. River runoff covers all water arriving in the hydrological network directly during rainfall or snowfall plus groundwater from the upper waterways feeding rivers throughout the year more or less evenly. Part of the groundwater, known as renewable water, does not join rivers, but flows to the sea or ocean or is spent on evaporation. In this situation, renewable water resources estimation utilizing river fluid data only yields underestimated findings. So it is of very practical importance to provide accurate data on the amounts of groundwater that cannot reach the river system. The regions with low-development hydrological networks, in specific plains with dry and semi-arid climates, are of the most significance. In such situations, groundwater resources can play an essential part in contributing to the overall volume of renewable water resources and the exhaustion happens after the rainy periods. In these situations, the reliable assessment of this groundwater in all locations of the world is a difficult subject due to the lack of critical data.

This estimate is nonetheless made for a number of areas and countries and permits for some inferences to be taken on a worldwide basis for diverse physiographical situations. The FAO made the most comprehensive assessment in 1995 for renewable sources of water, including rivers and unconnected groundwater. It is available in all African countries where more than half the continent is occupied by arid and semi-arid regions. According to the FAO data, for the entire continent, the total volume of renewable groundwater resources is 188 km^3 per year, which may amount up to 5% of the river runoff volume (Araujo et al., 2015). These principles are extremely significant and should be taken into consideration in particular countries such as Egypt, Libya, Tunisia, and Morocco in dry regions.

2.3. GLOBAL WATER RESOURCES

Water resources are usable or potentially helpful water supplies for people. Water uses include farming, industrial, residential, leisure, and environmental operations. Virtually everyone in this world needs fresh water. 97% of earth water is salty, with only 3% of it locked into glaciers and polar ice caps, which are only a little over two-thirds of freshwater. The rest of the freshwater source are found primarily as groundwater; just a small portion is present in the air and above ground.

The world supply of pure freshwater is constantly decreasing, whereas the freshwater supply is a renewable resource. The need for water in many regions of the world is already higher than supply and the world population continues to climb. It was recently that more than half of the world's wetlands and their vital environment services were destroyed in the 21st century to understand the worldwide relevance of maintaining water for ecosystem services. Freshwater environments rich in biodiversity are quicker disappearing than marine habitats or land ones. Water rights (when a framework exists) are known as the framework for distributing water resources for consumers of water.

2.3.1. Sources of Freshwater

2.3.1.1. Surface Water

Superficial water is a river, lake, or marsh of freshwater. Surface water is naturally refilled by rainfall and lost gradually by ocean outflow, evaporation, and under-surface sewage.

Watershed after precipitation is the only natural input in every surface water system, it depends on numerous other elements at all times as well. These include lake, wetland retention, and artificial reservoir capabilities, the porosity of the soil below these storage compartments, the runoff features of the watershed area, precipitation scheduling, and the levels of localized evaporation. All of these variables also affect the proportions of water lost.

Human activity can influence these elements extensively and sometimes devastatingly. Human beings often build reservoirs and reduce storage capacity by draining wetlands. People often increase runoff volumes and speeds by paving regions and channel flux.

The annual runoff of surface water can be quantified. This covers the quantity of rain drainage and snowfall left after nature is taken in, land evaporation, and vegetation transpiration.

The California Water Science Center tracks surface water flux and yearly runoff in locations such as California using a network of around 500 stream measurements which data are collected in real-time from across the state. Thus this helps to achieve the 8,000 stream measuring stations monitored by the USGS national stream measuring record.

In turn, documents and records of water data have been provided through the years. Management teams supervising water distribution can then decide on enough water supply for the sectors. These include municipal, industrial, agricultural, renewable, and reservoir storage (Badham et al., 2019).

A crucial aspect is the overall amount of water accessible at any one time.

Surface water, such as rivers, streams, and marshland, is water on the earth's surface. This can be called blue water as well. Precipitation and runovers from the neighboring areas produce the overwhelming bulk.

In spring, as the climate heats up, the snow melts down into neighboring rivers and streams that contribute to a big part of our water. Surface water levels are decreased as the water is evaporated, and water enters the subsurface and becomes groundwater

Surface water is also utilized for irrigation, treatment of wastewater, livestock, industrial applications, and recreation as well as for drinking. Approximately 68% of the water supplied to the municipalities comes from surface water as documented by the Environmental Protection Agency

(EPA). Surface water is termed freshwater for USGS Water Use reports when fewer than 1000 mg/L (mg/L) of solids are dispersed (Ali et al., 2019).

Three main kinds surface water exist. The surface waters are permanent (perennial) throughout the year. Rivers bogs and lakes are part of this. Half-string (ephemeral) surface water is water bodies that occur in locations such as rivers, lagoons, and waterholes only at particular times of the year. Surface water generated by people is water that can be perpetuated through technology created by humans.

These would be lakes, dams, and man-made marshes. The surface water contained in the dams could be used as hydropower for sustainable energy. Water is the obligation to generate energy in surface water generated by rivers and streams.

2.3.1.2. Under River Flow

The total amount of water moved downstream over the river is typically a mixture of apparent free water flow and significant contribution to the sub-surface rock and gravel underneath the river and its floodplain, the so-called hyporheic region.

This invisible part of the flow can substantially surpass the visible flow of many rivers in big valleys. The hyporheic zone frequently provides a dynamic interaction between surface water and real groundwater, which receives water from groundwater when water sources are completely loaded. This is particularly important in karst areas with common pot-holes and subsurface rivers.

The silt and porous area, under and next to a stream bed where superficial groundwater and surface water are combined, is a hyporheic zone.

The mechanics and behavior of flow in this area are considered vital for the interaction of surface water/surface water as well as fish breeding. The hyporheic zone can be built by engineers and maintained continuously for improving both the water quality and riparian ecology as a novel urban water management approach.

The composition of species inhabiting this area is called hyporheos. In 1959 Traian Orghidan invented the term hyporheic by merging Greek terms: hypo (below) and rheos (flow).

2.3.1.3. Hyporheic Zone and Hydrology

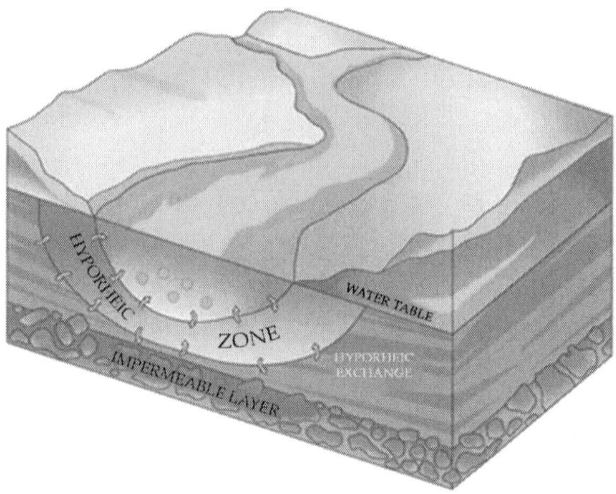

Figure 2.7. Cross-sectional diagram showing Hyporheic zone and hydrology.

Source: https://upload.wikimedia.org/wikipedia/commons/a/a3/Zona_iporreica_processo.gif.

2.3.1.4. Hyporheic Zone Process

The zone of hyporheic interchange involves the movement of water into and out of the stream bed and the transportation of dissolved gases and solutes, pollutants, microbes, and particles thereof. The hyporheic area may be merely several centimeters deep, or up to tens of meters deep or horizontally based on the underlying geology and geography.

A hydrologic study involves the theoretical concept of the hyporheic zone as a combination and storage zone. The first major notion for the hyporheic area is the time of stay; the water in the canal is moving at a considerably quicker rate than the hyporheic area, so this slower-water stream substantially boosts the period of water residence in the stream. Times of water residency affect the digestion rate of nutrients and carbon. Longer residence times encourage dissolved solvent storage that can subsequently retrieve the signals generated by the channel into the canal, postpone or alleviate them (Altenburger et al., 2015).

The other crucial idea is that of the interchange of hyporhea, or the speed of water coming into the subsurface or leaving that area. Water flows momentarily into the hyporheic area, but the water flows ultimately into the surface canal or serves to the storage. The efficacy of the hyporheic exchange is affected by fluvial structures with less streaming rugged water flow channels. Geomorphic features including stream meander patterns, bag sequences, huge woody debris dams, and other characteristics are driven to longer flow routes. The hyporheic area and its interconnections have an impact on the volume of water flowing downstream. Gains in reach show that the floodwater is released into the stream as water goes downstream, which boosts the volume of water on the primary channel. Paradoxically, the rippling of the stream is regarded "loosing" water when water penetrates into the groundwater zone leading to a significant loss of surface water.

The hyporheic area has a variety of advantages, such as:

- Habit and refuge for different species of fish, aquatic plants, and interstitial organisms.
- Reduction in pollutant concentrations dissolved on the water supply;.
- Water control and solvent exchanges between the main water supply and groundwater.

2.3.1.5. Groundwater

Freshwater can be found in the crevice of soil and rocks, the sub-surface water, or groundwater. Water flows also under the water table inside the aquifers. In a water tank (also dubbed "fossil water"), it may be beneficial to distinguish between sub-surface water that is intimately related to the sub-surface water of a water tank.

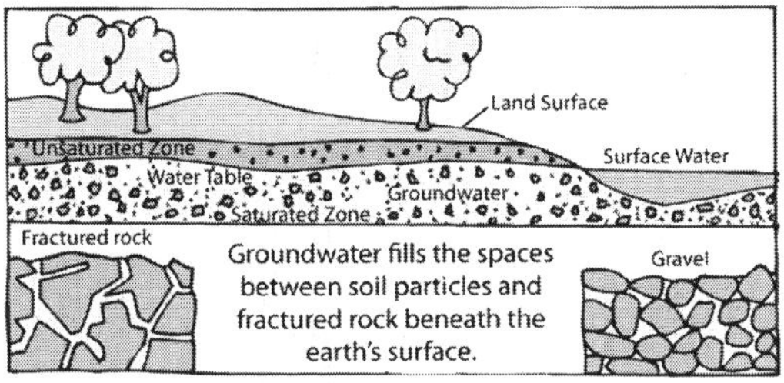

Figure 2.8. Groundwater is the water found underground in the cracks and spaces in soil, sand, and rock. It is stored in and moves slowly through geologic formations of soil, sand, and rocks called aquifers.

Source: https://www.groundwater.org/get-informed/basics/groundwater.html.

Inputs, outputs, and storage can be considered in the same way as surface water. Because of its sluggish settlement, sub-surface water storage is often substantially bigger than surface water, as opposed to inputs. This distinction facilitates the long-term unsustainable use of sub-surface water by people without serious implications However, the average rate of filtering above an under-surface water source, in the long run, is the upper limit of average water intake (Ashton & Turton, 2009).

Groundwater is the water found in rock and soil pore spaces below the earth's surface and rock compositions fissures. An aquifer is in reality a unit of rock or an unconsolidated deposit of the useable amount of water that can be collected and used.

The depth at which soil pores are totally waterlogged is called the water table, gaps or fractures, and voids of rock. Groundwater is refilled from the surface; it can naturally release at springs and springs from the top layer, and it can provide oasis or humidity. Furthermore, groundwater is often extracted through the construction and operation of pools for agricultural, municipal, and industrial usage. Hydrogeology, commonly termed groundwater hydrology, is the study of the circulation and flow of groundwater.

Groundwater is typically considered to flow into shallow aquifers but can also, in a technical way, contain soil humidity, permafrost (frozen soil), stationary soil water, and deep geothermal and oil-built water at very low permeability. Groundwater is considered to offer lubrication which may affect the movement of defects. Most of the earth's subsurface probably contains water that in some situations can be blended with other fluids.

Groundwater is frequently less expensive and handy than surface water and less sensitive to pollution. It is therefore widely utilized for the provision of public water. Groundwater, for instance, is the biggest source of potable water supply in the US, while California pulls the most quantity of groundwater in the United States every year. Underground reservoirs hold significantly more water than all the reservoirs and lakes in the United States, including the Great Lakes. Much of the country's water resources come from groundwater alone.

Groundwater use has connected problems for the environment. For example, polluted groundwater in rivers and lakes is less obvious and harder to remediate than pollution elsewhere. Pollution of soil water arises mostly from unsatisfactory land disposal of garbage. Main resources usually involve industrial and domestic chemicals and landfills, agricultural surplus fertilizers and pesticides, industrial waste lagoons, tailings, and process wastewater, industrial frackings, oil field salt pitting, underground oil storage tanks and pipelines, and septic systems (Zhao et al., 2017).

Furthermore, in coastal places, groundwater is prone to saltwater infiltration and can induce land subsidence if drained in an unsustainable manner, resulting in sinking cities (such as Bangkok) and altitude loss (such as the multiple meters lost in the Central Valley of California). Sea level increase and other factors caused by climate change, which will alter precipitation and water shortage globally, exacerbate these challenges.

2.3.1.6. Characteristics and Location (Aquifers)

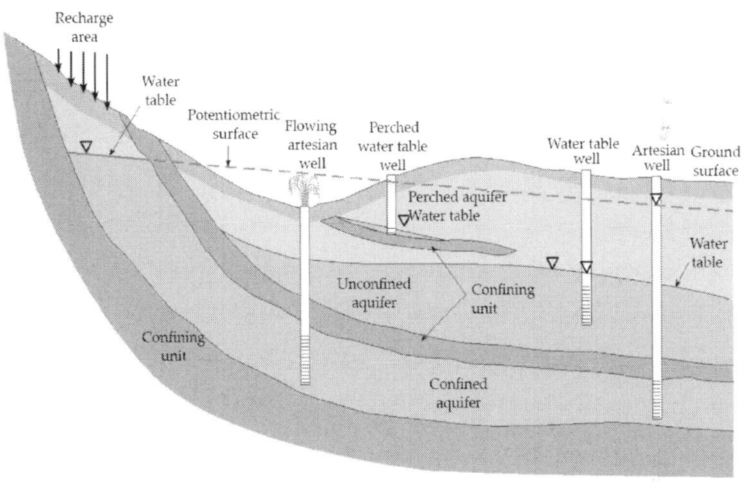

Figure 2.9. One big distinction from the non-confined aquifer is that the saturated zone is not dewatered by gravity discharge while an Artesian aquifer is being pumped. A well pitching an unconfined aquifer over a confined aquifer can drain the former through gravity drainage and does not influence the artesian aquifer, assuming its containment bed is insignificant.

Source: https://www.ngwa.org/what-is-groundwater/About-groundwater/confined-or-artesian-groundwater.

An aquifer is a permeable layer of soil that holds and transfers groundwater. The aquifer is unconstrained when water can flow freely in between the surface and the saturation zone. Gravity forces water to flow down; therefore, the deeper areas of unconfined aquifers are typically more filled.

The water table, also known as the phreatic surface, is the top layer of a saturated layer in an aquifer system. The phreatic zone is located just below the water table, in which all pore spaces are waterlogged. An aquitard is a low-porosity substrate that allows for modest groundwater transport. An aquiclude is a substrate with porosity that is so low it is virtually impermeable to groundwater.

An aquiclude is a base with extremely low porosity, making it essentially impervious to groundwater. An aquifer that is constrained is covered by a somewhat impermeable layer of rock or substrate, as in an aquiclude or aquitard. Groundwater can get pressured as it runs if a confined aquifer has a lower grade from its recharge zone (Vyas & Nath, 2021). This can result in artesian wells that flow freely without the use of a machine and ascend to a higher height than the unconfined aquifer's static water level.

A restricted aquifer carrying groundwater under positive pressure is known as an artesian aquifer. Water is confined in an artesian aquifer, which is encircled by strata of impermeable rock or clay that create a positive force on the water inside the aquifer. Water in the well-pipe would ascend to a height comparable to the point where hydrostatic balance is achieved if a well were sunk into an artesian aquifer. An artesian well is constructed into such an aquifer. A running artesian well is one in which water hits the ground surface underneath the aquifer's natural pressure.

Many recently discovered oil wells are pressurized, therefore fossil water aquifers can be artesian if they are under enough pressure from the adjoining rocks. It can be deduced from the preceding remark that not all aquifers are artesian (i.e., water table aquifers occur where the groundwater level at the top of the aquifer is at equilibrium with atmospheric pressure). When the water table in an aquifer's recharge zone is above the well's head, the aquifer has the recharging ability.

2.3.1.7. Water Cycle

Groundwater accounts for nearly a one-third of the world's freshwater supply or 0.76% of all water on the planet, counting oceans and permanent ice. Groundwater makes up nearly all of the world's liquid freshwater. The entire amount of freshwater held in the snow and ice pack, as well as the north and south poles, is about equivalent to global groundwater storage. This makes it a valuable resource that can operate as a natural storage tank, buffering against surface water scarcity such as those experienced during droughts.

Figure 2.10. Dzherelo, a common source of drinking water in a Ukrainian village.

Source: https://en.wikipedia.org/wiki/File:Shipot.jpg.

The surface water from precipitation, streams, and rivers gradually replenishes groundwater as it enters the water table. In contrast to the short-term stores of water, such as atmosphere or fresh surface water, groundwater can be a long-term reservoir for the ecological cycle of water (during periods between days and entire millennia). The figure below demonstrates how long it can take to finish its normal cycle for deep groundwater (which is fairly remote from the surface recharge).

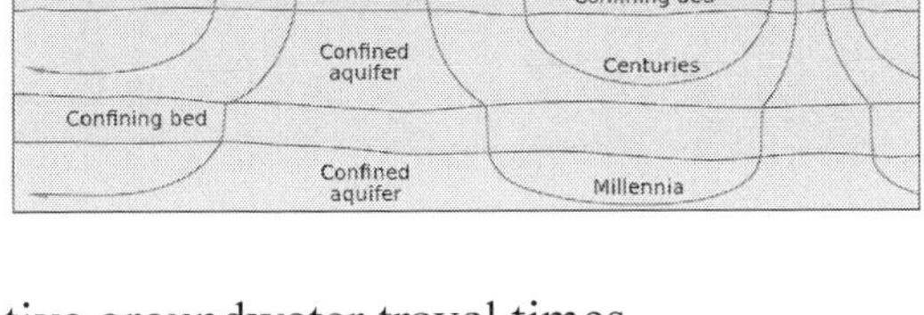

Figure 2.11. Relative groundwater travel times.

Source: https://en.wikipedia.org/wiki/File:Shipot.jpg.

The Great Artesian Bay is one of the biggest restricted aquifer systems in the world in central and eastern Australia with a surface area of over 2 million km^2. Hydro-geologists have indeed been able to identify water collected from these aquifers can be over 1 million years old by analyzing the trace components of water produced from deep beneath (van Koppen & Schreiner, 2014).

In evaluating the groundwater age derived from different regions of the Great Artesian Basin, the age variations across the basin were identified by hydro-geologists. The ages are younger when the water recharges the water on the eastern side. When groundwater moves all over the country to the west, it gets older and is in the west the oldest groundwater. This implies the groundwater running through the Great Artesian Basin moves at an estimated rate of roughly 1 m per year, to nearly 1000 km from recharge sources over 1 million years.

Figure 2.12. Reflective carpet trapping soil water vapor.

Source: https://en.wikipedia.org/wiki/Water_resources#Natural_sources_of_fresh_water.

A recent study has shown that groundwater evaporation can serve a major role in the local water cycle, particularly in arid areas. In Saudi Arabia, scientists considered recapturing and recycling this evaporating rainwater for agriculture irrigation. In a five-month dry, plant-free desert environment without rain or irrigation the photo above shows a 50-cent square reflecting carpet created out of little, neighboring plastic cones. It was managed, with a green region of roughly 10% of the area of the tapestry, to catch and condense sufficient ground vapor to bring naturally buried seed. It is implied that, if seeds were planted before laying this carpet, a much bigger area would become green.

2.3.1.8. Temperature

The high specific water thermal capacity and the solar and rock insulation effects can minimize the climate effects and keep a comparatively stable temperature in groundwater. In some regions, where groundwater temperature remain at around 10°C (50°F), groundwater could be used for temperature control within surface constructions (Singh et al., 2021). For example, relatively cool groundwater can be drawn into a home by radiators and then released to the ground in yet another well during warmer weather. Because water is generally warm during the cold seasons, water may be used considerably more efficiently as a heat source for heat pumps rather than air.

Quantities

Groundwater volume can be determined on an aquifer through the measurement of water levels in local wells and geological data from well-boiling, to establish the range, depth, and sizes of waterborne sediments and rocks. In order to evaluate the depths of water, and to gather sample soil, rock, and water for lab analysis, test wells may be explored before the money is invested in production wells. Pumping studies in test wells can be carried out to assess the aquifer fluid properties.

Aquifers have different features with the geology and structure of the substratum and its topography. In general, sedimentary geological formations include more productive aquifers. In many situations, compared to worn and cracked crystalline rocks, the soil yields fewer quantities. The most effective sources of groundwater are unconsolidated to weakly solidified alluvial materials accumulating as valley filling deposits in large river valleys and geological subsidizing basins.

The flows of fluid can be changed by brittle rock deformation in fault zones in diverse lithological contexts; fault zone hydrogeology is the science of the processes by which this happens.

2.3.1.9. Frozen Water

Runoff of glaciers is termed surface water. Several strategies to use icebergs as a supply of water have been suggested; however, this is still done for scientific purposes.

The Himalayas frequently referred to as 'The Roof of the World,' features some of the largest and roughest highlands in the world, along with the largest glacier and permafrost zones outside the poles. The water from the Himalayas flow to ten of Asia's main rivers, and over a billion people rely on them. Temperatures rise faster than the worldwide norm to complicate problems.

In Nepal, in the last decade the temperature has increased by 0.6°C while worldwide, over the last century, the Earth warmed about 0.7°C.

Figure 2.13. Iceberg near Newfoundland.

Source: https://en.wikipedia.org/wiki/Water_resources#/media/File:Carlb-newfoundland-iceberg-2002.jpg.

2.4. ARTIFICIAL SOURCES OF USABLE WATER

Treatments of wastewater (reclaimed water) and desalinated saltwater include artificial sources of freshwater.

2.4.1. Reclaimed Water

The process of turning wastewater into the water, which may be reused for various purposes, is called water recycling (also known as wastewater reuse). Irrigation or refilling of surface water and groundwater might involve the re-use of gardens and agricultural areas (i.e., groundwater recharge). The reused water can also be used in houses, businesses, and industries to meet certain requirements (e.g., flushing toilets). It can even be processed to meet the requirements for drinking water. The latter alternative, depending on the methodology utilized, is either named "Direct Drinking Reuse" or "Indirect Drinking Reuse."

Instead of using freshwater, recycling of wastewater can be a water-saving strategy. If consumed water is ultimately returned to natural sources of water, ecosystems can still gain as part of the natural cycle of water, boosting stream flows, supporting plant life, and recharging aquifers.

Wastewater reuse is a long-standing irrigation strategy, particularly in dry regions. Treated wastewater can remain an alternate water supply for human activities when reused as part of sustainable water management. Shortages can be reduced and demands on groundwater and other natural pools of water are reduced. The nutritional content of the effluent, which can lessen the demand for other fertilizers, is another possible good element (Singh et al., 2020).

The frequently claimed drawbacks or concerns of recycled wastewater include contents of potentially dangerous elements such microorganisms, heavy metals, or toxic organic contaminants (including pharmaceuticals, personal care products, and pesticides). Irrigation with wastewater could, based on the characteristics of wastewater and on-ground or plant properties, have good and negative impacts on soil and plants.

Developing more sustainable sanitary and wastewater management will necessitate a focus on measures related to the management of resources such as wastewater reuse or excrete reuse that keeps valuable assets for constructive use. In turn, this promotes human welfare and broader sustainable development.

The recovered water is simply used more than once before it is returned to the natural flow of water. Advances in wastewater technology enable communities to reuse water for a variety of reasons. According to the water source and use and the way it is being supplied, the water is treated differently.

Every water on earth cycles again and again through the planetary hydrosphere, but the terms "recycled water" or "recovered water" typically means wastewater from the home or the enterprise to a wastewater treatment facility through a sewage system, which is processed at a level in accord with its intended application.

The WHO has highlighted the following main driving points for the reuse of wastewater:

- Water shortage and stress increase.
- Growing populations and food safety challenges associated with this.

- Rising pollution from inappropriate dumping of wastewater.
- Growing recognition of the waste, excreta, and greywater resource value.

Water recycling and reuse, not just in the arid areas but also in cities and contaminated surroundings is increasingly important. The groundwater aquifers utilized by more than half of the world's population are over-drafted. Reuse is continue to grow with the increasing urbanization and concentration of the world's population along the coastlines that have limited or big capital supply.

The reuse and recycling of wastewater, reduction of environmental contamination, and improvement of carbon footprint can reduce large volumes of freshwater. The alternate water supply option can be reuse.

Making untreated wastewater halfway and significantly boosting recycling and safe reuse worldwide by 2030, in accordance with United Nations Sustainable Development Goal 6.

2.4.2. Desalination

The technique of desalination removes components of minerals from salty water. More generally, desalination is a problem for agriculture in removing salts and minerals from a target substance, for example in soil desalination. Saltwater is desalinated to generate water for human consumption or irrigation.

Salt and brine are the by-product of the procedure of desalination. Desalination is utilized on many sea and underwater ships. Most of the modern desalination attention is centered on the economic supply of freshwater for human use. It is one of the few rainfall-independent water supplies, together with recovered wastewater (Singh et al., 2021).

As a result of its energy consumption, desalination of marine water is sometimes more expensive than the surface and groundwater freshwater, recycling water, and water retention.

But they are not always accessible and depletion of resources is a major problem globally. Desalination processes are commonly powered by either thermal (for distillation) or mechanical (for reverse osmosis) as primary energy.

2.4.2.1. Applications

In 177 countries, there are around 16,000 operating desalination plants generating around 95 million m3/day freshwater. Desalination presently accounts for approximately 1% of drinking water in the globe. In countries in the Middle East and North Africa region, like Saudi Arabia, the UAE, and Kuwait, desalination is very popular. Kuwait generates more water than any other country by desalination. Desalination in small island and developing countries is also an essential source of water.

Some 21,000 desalination units throughout the world are now operational. The United Arab Emirates, Saudi Arabia, and Israel are largest the countries performing desalination. Saudi Arabia (Ras Al Khair) with a volume of 1,401,000 cubic meters per day, provides the largest desalination factory in the world (Sroufe, 2018).

Desalination is costly relative to most alternative water sources and desalination only satisfies a very small portion of human use. It is usually only economically viable in dry locations for highly valued applications (e.g., residential and industrial applications). Desalination for agricultural use, however, is growing in locations such as Singapore and California which are highly populated. The Persian Gulf is where desalination is the most widely used.

While the costs are reducing, and the technology for wealthy locations in the vicinity of the oceans usually is generally good, a 2004 research argued that "Desalinated water may be a solution to some water-stress zones, but not too low, deep inland or high elevations. Unfortunately, this includes some of the places with major water difficulties," and, "indeed, it is necessary for transporting the water by 2000 m or over 1600 km to be equal to the price of desalination. Thus, transporting fresh water from somewhere else may be more cost-effective than desalination. Transportation expenses could mirror desalination expenses in places far away from the seas like New Delhi or in areas as expensive as Mexico City. Desalination is also expensive in places, both a little distant from the sea and quite tall, like Riyadh and Harare. In comparison, transportation costs are far lower in other destinations locations like Beijing, Bangkok, Zaragoza, Phoenix and obviously, seaside towns like Tripoli." Saudi Arabia is piped into Riyadh 320 kilometers inland after desalination in Jubal. Desalination is becoming a competitive choice for coastal towns.

Figure 2.14. The largest reverse osmosis water desalination plant in the world so far constructed is the Ras Al-Khair Power and Desalination Plant located in Saudi Arabia which produces over 1 million cubic meters of water a day. The construction of the desalination plant was completed in 2014 and delivers about 1 million cubic meters of freshwater a day along with 2400MW of electricity that is generated at the plant.

Source: https://constructionreviewonline.com/biggest-projects/largest-reverse-osmosis-water-desalination-plant/,

Desalination is an artificial process through which salt water is turned into freshwater (usually seawater). Distillation and reverse osmosis are the most popular desalination methods.

Multiple techniques exist. Each one has advantages and inconveniences, but they are all useful. The processes of desalination are separated into membrane-based methods of reverse osmosis and thermal techniques of distillation (i.e., multi-stage flash). Desalination is traditionally done by boiling and re-condensing seawater to leave salt and contaminants behind.

There are two main technologies worldwide, MFD, and reverse osmosis. There are two more desalination capacities.

In conclusion, for all life on the earth, the water is a crucial resource. Only 2.5% of earth's water resources are fresh. In ice caps and glaciers are two-thirds of the freshwater. Of the remaining one percent, one-fifth is not easily exploited in remote, inaccessible locations and seasonal precipitation in moon-tree flooding. Water becomes scarcer with time; access to clean, safe, and potable water increasingly becoming unavailable in many countries. Today, humanity exploits just approximately 0.08% of the world's fresh water in the ever-increasing need for sanitation, drinking, production, leisure, and agriculture. Because of the limited proportion of water left over, there has been constant difficulty to optimize the freshwater left from natural resources across the globe (Afroz et al., 2014).

A great deal of effort is being made to optimize water consumption and to reduce the environmental impact on the natural environment of water use. The observation of water as an essential part of the ecosystem is based on holistic management of water resources that helps in determining the amount and quality of the environment.

Water supply is facing problem as it is a scarce resource. This was taken into account by a project enacted by the European Union Seventh Framework Program for Research, Technological Development and Demonstration, DESAFIO (the acronym for Democratization of Water and Health Management through Means of Socio-Technical Innovation). This project was confronted with a difficult problem to develop: to eliminate socioeconomic structural inequities in the access to water and public health services. The DESAFIO engineers are working on a solar-powered water treatment system that supplies potable water to a very impoverished village in Minas Gerais.

Successful resource planning requires intimate knowledge of resources available, the use of resources available, contending resource requirements, measures for and procedures for assessing the importance and value of competing requirements, and mechanisms for translating policy decisions on a field basis into action. Overall, new managers have to develop new experiences and share experiences with foreigners to promote the government to implement these new policies.

This is especially problematic for the development of water as a resource since water sources can transcend numerous national borders, and water uses contain problems for financial value assignment and are also difficult to manage conventionally. For example, uncommon or long-term values of old groundwater reserves, often called fossil water, or ecosystems.

CHAPTER 3

THE SCARCITY OF CLEAN WATER

CONTENTS

3.1. INTRODUCTION

Water is essential for life and livelihoods, as well as for long-term development. Many of the 17 Sustainable Development Goals (SDGs) will be achieved as a result of successful water management, including SDG 6: "Ensure availability and sustainable management of water and sanitation for all." Despite this, water seems to have become a pressing socioeconomic and geopolitical issue, and it is already a major national worry in some areas. By 2035, up to 40% of the world's population will be living in severely water-stressed places, and ecosystems' ability to produce fresh water would be severely harmed (Afshar et al., 2015).

River basins that span national borders provide 60% of fresh water. Trans boundary water agreements must be strong enough to deal with more unstable climate and soil conditions, as well as sociodemographic developments that will double the world's population to 9.7 billion by 2050 and quadruple the number of people living in cities. Conflict can and has arisen as a result of differing perspectives about water. Water as a human right and a shared public and environmental good is frequently countered by the idea of water as a commodity that must be priced to ensure effective and sustainable use. To enable peaceful and effective integrated water resource management and sustainable use, not just nations, but provinces and communities will need to harmonize their water viewpoints.

Figure 3.1. The direct effect of water shortages on the health of the continent in Africa is most obvious. Water-borne conditions like typhoid fever, cholera, dysentery, and diarrhea are frequent since individuals are compelled to drink clean water from questionable sources. Water shortages force many individuals to conserve water in their homes, increasing the danger of contamination of home water and malaria occurrences and mosquito-borne dengue.

Source: https://goinggreenrecyclingnigeria.org/clean-water-scarcity-in-sub-saharan-africa/.

Effective management means addressing ignored problems, like wastewater in current systems, projected at up to 30%; common institutional dysfunction, illegal conduct, low accountability and bribery in many countrie's water sectors.

In this chapter, the imminent water crises are brought to the fore in six inter-connected contexts; water shortage and insecurity; disasters related to water; crisis in water, sanitation and health (WASH). To counteract these grim water trends, UN agencies, governments, and civil society organizations have all stated that radical new approaches to water are required. Water will continue to support life, growth, and biodiversity for our coming generations only if we address these issues thoughtfully and collaboratively.

Water is abundant, but only 2.5% of all water is freshwater, even though it covers 70% of the earth. By 2050, this finite resource will have to support a projected population of 9.7 billion people, with an estimated 3.9 billion people – or more than 40% of the global population – living in highly water-stressed river basins. Water resources are being strained by more than simply population growth. Excessive consumption is also evident: during the twentieth century, the global population tripled, but water use climbed six-fold. Water demands are predicted to rise by 400% from industries and 130% from residential consumption between now and 2050 (Prinz & Singh, 2000).

The geographic and temporal imbalance between fresh water demand and supply is at the heart of worldwide water shortages. The main driving causes for the expanding worldwide demand for water are the growing global population, better living standards, changing consumption patterns, and the expansion of irrigated agriculture. Climate change-related factors like changing weather patterns (especially droughts and floods), deforestation, rising pollution greenhouse gas emissions, and inefficient water use can all contribute to a shortage of water.

Scarcity fluctuates over occur as a consequence of natural hydrological fluctuation, but it fluctuates much more as a result of current economic policy, planning, and management practices. Scarcity is likely to worsen as most forms of economic development progress, yet many of its causes can be predicted, avoided, or alleviated if they are correctly identified.

According to the United Nations' International Resource Panel, governments have a history of investing substantially in inefficient solutions such as dams, canals, sewage systems, pipelines, and water reservoirs, which are neither environmentally nor economically feasible. As per the scientific panel the most cost-effective method of cutting water use off economic growth is for governments to develop holistic water conservation plans that take the entire water cycle into account: from source to distribution, economical use, processing, recycling, reusing and returning to the environment.

Figure 3.2. One of the top causes of death among children fewer than five are water-related illnesses, inadequate hygiene and sanitation in Niger. In Niger, there are relatively low levels of access to drinking water and sanitation with great inequalities between cities, rural areas and regions.

Source: https://www.unicef.org/niger/water-sanitation-and-hygiene.

As the availability of water decreases, competition for this finite resource will intensify. There are an estimated 592 trans boundary aquifers, and 60% of all surface fresh water comes from internationally shared river basins. Water availability for human, economic, and environmental demands depends on continued international collaboration and coordination. Hundreds of international water accords have been negotiated through time, but it is not always obvious how countries will cooperate to manage growing resource strains so that they do not lead to further water wars.

Drought can worsen water insecurity. Drought affects more people than any other natural calamity. In 2016, 411 million people were affected by disasters, with drought affecting 94% of those affected. Droughts are also the most expensive disasters, wreaking havoc on agriculture in particular; droughts cost the US agriculture industry an average of US$6–8 billion a year (Partha et al., 2019). Drought has led to an annual loss of almost 27 million tons of grain output in China in the previous two decades, and the average annual drought area between 1950s and the start of this century has increased, from 11,6 million hectares and annual growth of 116%, to 25,1 million hectares. The prospective global welfare income to decrease risk would have been US US$94 billion in 2010 if water is supplied to irrigate agriculture. Findings also demonstrate that increased water safety can contribute to stabilizing production and prices for crops. Under water-safe conditions, worldwide production of wheat has decreased from 83% to 38% from below 650 million tons per year. It is crucial to prevent shocks and stressors by investing in water infrastructure. It calculated that in 2009 the World Bank would be at high and medium risk of vulnerability to climate change by 2030, around half of the Bank's water portfolio – then committed at 8.8 billion dollars and pipelines for 11.30 billion dollars. Furthermore, hydrological dangers lead to serious death, displacement and injury. In the past two decades, floods are the most frequent global natural disaster, 50% of all recorded events related to flooding in 2016. The most frequent occurrences of all disasters are related to water. The entire value of all flood-prone assets is projected to amount to US$45 trillion in 2050 – an increase of almost 340% compared to 2010.

Figure 3.3. This appalling data must be made available to the public to highlight the severity of the water shortages both locally and globally. Not only are those areas that are generally dry influenced by every continent throughout the planet. For at least one month a year, at least two billion individuals are affected. There is no access to clean drinking or drinking water for more than 1 billion people. This addition contains the consequences of water shortages and what it means to be without them.

Source: https://www.conserve-energy-future.com/causes-effects-solutions-of-water-shortage.php.

In 1970–2010, from EUR 3.7 billion to EUR 6.9 billion, the world's population grew by 87%. The average yearly population exposed to flooding increased from 33.3 million to 70.4 million per year by 112% over that period. The number of people susceptible to flood disasters would rise to 2 billion by 2050, due to increasing populations in areas prone to flooding, climate change, deforestation, wetland losses and rising sea levels.

Since hundreds died in Guatemala, the US and southern France in 2015, the UN was encouraged to broadcast urban flash warnings – noting that these catastrophes have become more prevalent in the past 2 decades under the change of the climate and intensive flooding and urbanization. Ecosystems associated with water can reduce calamities associated with water. In 2012, more than 625 million dollars in Hurricane Sandy damage had been evaded, for every hectare of mangrove and coastal marsh, worth $15,161 a year in disaster-related services (Owusu et al., 2016). Spending a million US dollars a year for reefs in Folkestone Marine Park on the west coast of Barbados, as an example of their efficiency in the risk reduced could reduce yearly storm losses in reefs by 20 million US dollars in this area. Despite these significant reductions in risk, ecosystems linked to water are decreasing globally. Up to half of the coastal ecosystems of the mangroves were degraded or eliminated in regions of Asia and America.

3.2. TERMINOLOGIES

3.2.1. Physical Water Scarcity

Physical water shortage occurs when there is insufficient water to satisfy all needs, including those necessary for ecosystem function. Physical water scarcity is a common occurrence in arid climates. It can also happen when water is abundant but resources are over-committed, such as when irrigation infrastructure is overdeveloped. Environmental deterioration and diminishing groundwater levels are two signs of physical water scarcity. Because all living things require water to survive, water stress is harmful to them.

Physical water shortage occurs when natural water supplies are insufficient to meet a region's demand, while economic water shortage occurs when adequate water supplies are mismanaged. As per the United Nations Development Program, the latter is more frequently considered to be the cause of countries or regions experiencing water scarcity, as most countries or regions get enough water to cover domestic, industrial, food production, and environmental needs, but lack the infrastructure to make it accessible (Olmstead, 2014). Around one-fifth of the global population now lives in areas affected by physical water shortage, which occurs when a country's or region's water supplies are insufficient to meet demand, including the water required for ecosystems to function successfully.

3.2.2. Economic Water Scarcity

Economic water scarcity results from a lack of infrastructure development or technology to extract water from rivers, reservoirs, or other water sources, as well as a lack of human ability to meet the water requirements. Economic water scarcity affects one-quarter of the world's population. Economic water shortage is exacerbated by a lack of infrastructure, forcing people without regular access to water to travel great distances to obtain contaminated water from rivers for home and agricultural use.

3.2.3. Water Crisis

Humans, like many other species, rely on water as a resource. When a country's population grows, so does its water demand. According to UN estimates, by 2040, over 4.5 billion people will be afflicted by a water crisis. Furthermore, as the population grows, there will be a demand for food, and for food output to keep pace with population expansion, there will be greater demand for water to irrigate fields. As the population grows, so does the water demand, resulting in a water crisis in which there is insufficient water to share at healthy levels.

Figure 3.4. There are some rural regions in the whole world dealing with water shortages because they're just around the corner. Places deemed desert or remote areas cannot have sources in which humans can efficiently obtain water.

Source: https://www.conserve-energy-future.com/causes-effects-solutions-of-water-scarcity.php.

3.2.4. Indicators and Water Stress

Hydrologists now usually evaluate water shortages by examining the population-water equation. This is accomplished by comparing the total water resources available to the population of a nation or region every year. A typical technique for water shortage measurement was to classify countries according to annual water resources per person. According to the Falkenmark Water Stress Indicator, a nation or region is considered to be under "water stress" when annual water resources fall below 1,700 cubic meters per person per year. Periodic or limited water shortages can be foreseen at levels between 1,700 and 1,000 cubic meters per person per year. The country experiences "water scarcity" when water supplies fall below 1,000 cubic meters per person per year (Olsson & Andersson, 2006).

3.2.5. Others

Examining the physical existence of water in nature and comparing nations with lower or larger volumes of water accessible for usage are two more ways to measure water shortage. This method frequently fails to represent the water resource's accessibility to the population who may use it. Others have linked population density to water availability.

3.3. WATER RESOURCES

3.3.1. Availability

The UN estimates that just 200,000 cubic kilometers (162.1 billion acre-feet) of freshwater is accessible for human use, with 1.4 billion cubic kilometers (1 quadrillion acre-feet) on the earth. On the earth, only 0.014% of the water is fresh and easy to reach. Of the remaining water, 97% are saline and fewer than 3% are hard to reach. A total of 14,000 cubic kilometers of freely available fresh water in the form of groundwater (in water, for instance) and surface water (rivers and lakes) (nearly 3359 cubic miles). Only 5,000 cubic kilometers of this entire amount are utilized and reused by humans. On a global basis, the amount of freshwater available is technically sufficient (Nazemi & Wheater, 2015). In principle, there is enough freshwater to supply the needs of the world's present population of over 7 billion people, and even to support a population increase to 9 billion or more. Water, nevertheless, is a limited resource in some regions of the world and for some segments of the population due to unequal geographical distribution and, in particular, unequal usage.

3.3.2. Renewable Freshwater Resources

When assessing water shortages, a statistic called renewable freshwater supply is frequently utilized. This statistic is useful since it may describe each country's entire available water supply. It is possible to determine whether a country is prone to physical water shortages by knowing the entire available water supply. Because it is an average, this statistic has flaws; precipitation distributes water unevenly around the globe each year, and yearly renewable water resources change from year to year. This statistic also does not account for individual, family, industrial, or government access to water. Finally, because this statistic describes an entire country, it does not adequately depict whether or not that country is facing water scarcity. Even though both Canada and Brazil have abundant water supplies, they still face a variety of water-related issues. Tropical countries in Asia and Africa have a scarcity of freshwater resources, as can be seen.

3.3.3. Demands

The amount of freshwater available on the planet is about 1% of the total amount of water in the world. Streams, groundwater, wetlands, lakes, glaciers and rivers, provide the majority of this water. Six out of every ten people are at risk of becoming water-stressed, as global temperatures and water demand are rising. Globally, wetlands are drying out at a rate of roughly 67%, putting a large number of people in danger of water stress. Water stress is expected to affect two-thirds of the world's population by 2025, as global demand for water and climate temperatures rise.

3.4. CONTRIBUTING FACTORS AND CAUSES

3.4.1. Depletion of Freshwater Resources

Other resources of freshwater, such as groundwater and glaciers, have been established as a supply of clean water. This is in addition to the typical surface waters, such as streams and lakes. Groundwater is water that pools underneath the earth's surface and can produce a useable amount of water through springs and pools. These locations are also known as aquifers where groundwater is gathered. Glaciers supply fresh water in the form of melted water or melted from snow or ice, which provides flows, or springs with increasing temperatures. The availability of traditional fuels is declining as a result of issues like pollution or disappearance due to climate change, are increasingly drawing on these sources. Human population growth is an important factor in the increased consumption of water resources of this kind.

3.4.1.1. Groundwater

Because of the growing human population, there is increasing competition for water that deepens many of the world's major aquifers. This is both owing to direct human consumption and irrigation by groundwater in agriculture. There are already millions of pumps of every size which draw groundwater worldwide. Irrigation is provided by groundwater and is extracted at unsustainable rates in dry places, such as India, Nepal, and North China. Shanghai, Madras, Beijing, Bangkok, and Mexico City, are among cities with waterfalls between 10 and 50 meters long.

Figure 3.5. The consumption of water necessary to support life of huge populations also increases with the world population. Water contributes to almost every area of human life, although the creation of food is the majority. More than 70% of the world's water use is made up of agriculture. In addition, 2000–3000 liters of water are necessary to generate adequate food for the daily consumption of calories for each individual (FAO).

Source: http://www.fides.org/en/news/62171-AFRICA_SUDAN_Serious_scarcity_of_clean_water_in_Port_Sudan_aggravates_plight_of_displaced_persons.

Groundwater has not been a widely-used resource until recent history. In the 1960s, groundwater aquifers became increasingly established. Changes in understanding, technology and financing have enabled the concentrated development of groundwater abstraction from surface water resources. These transformations have enabled social advancement, such as the "agricultural groundwater revolution" and the expansion of irrigation, to enhance food production and rural development. Nearly half of all the drinking water in the world comes from groundwater.

In most aquifers, there are vast quantities of water stored under the ground which have significant buffer capacity to drain water during periods of drought or low precipitation. This is vital for individuals living in areas that cannot rely solely on precipitation or surface water to ensure reliable availability of water throughout the year. By 2010, worldwide aggravated abstraction of groundwater has been calculated at over 1,000 km^3 per annum, with 67% irrigating, 22% home irrigation and 11% industrial use. 72% of all water use abstracts globally are among the top 10 main use of abstract water (Italy, Indonesia, Saudi Arabia, Mexico, Bangladesh, Iran, Pakistan, America, China, and India). The livelihoods of 1.2 billion to 1.5 billion rural households in the poorer countries of Africa and Asia have groundwater become vital for food security (Muller, 2015).

While groundwater sources are very common, the renewal rate or recharge rate of various groundwater sources is one of the main concerns. If extraction from non-renewable groundwater sources is not adequately controlled and managed, they can lead to depletion. The decreased water quality of the source over time is another worry for increasing soil water use. In groundwater systems there is usually a decrease in natural outflow, falling stored volumes, lowering water levels and water degradation. Exhaustion of groundwater may have various negative repercussions, such as higher cost of groundwater pumping, increased salinity and other changes in water quality, earth declines, and depleted springs and decreased base flows. This valuable resource is also hazardous to human pollution.

If a major facility is to be set up near an abundant area of water, water-borne firms must extract groundwater from a source at a pace higher than that resulting in an ongoing fall in the level of groundwater. The groundwater is removed, bottled and then shipped throughout the country or world, and this water is never returned. If the water table falls below a threshold level, the companies moving from the area will leave a serious scarcity of water (Alamanos et al., 2018). Everybody and everything in the area where the water is used will be affected by the depletion of groundwater: tourism, ecosystems, animals, businesses, farmers, and other consumers. The water table is drained evenly in the area with millions of gallons of water coming out of the ground and not only because the water table is connecting to the earth. Bottling plants produce water shortages and ecological equilibrium. They lead to droughts in water-stressed areas.

Figure 3.6. While all of us depend on water to survive, we also contribute to increasing water shortages. A multitude of causes, including various human activities, are endangering the vitality of rivers, lakes and other bodies of freshwater. Pollution, climate change, industrial agriculture, unsustainable energy generation and population increase are all among them. The eventual outcome is an increase in water scarcity around the planet.

Source: http://inwrdam.org.jo/2020/06/how-people-are-resolving-to-reduce-water-scarcity.

3.4.1.2. Glaciers

Approximately 2% of earth's water in the glaciers is frozen. These glaciers give freshwater in any location along the glacier's ripple as they melt naturally. Locals use this water for several factors including agriculture, animal hydropower and cattle. This helps to minimize water shortage, as a small number of individuals have more water available. Total glaciers around the world have been estimated to be 60% of what they are presently in 2100. Climate change is the primary cause of these glaciers melting (Ludwig et al., 2014). Glaciers reflect sunlight into space, resulting in a reduction in global temperatures. This is known as albedo, and if the glaciers didn't reflect sunlight, temperatures would gradually rise. Glaciers will melt faster as temperatures rise, limiting the total quantity of sunlight reflected globally. Melting glaciers begin to recede over time and will be tough to rebuild once seasonal changes take place. The loss of mass of glaciers, combined with retreating glaciers, may reduce yearly run-off, affecting water availability in many cold parts of the world. In some seasons, about a third of glaciers may see a 10% decline in run-off.

3.4.2. Expansion of Industrial and Agricultural Users

The intensive use of water in agriculture/livestock breeding and industry is the primary cause of scarcity as a result of consumption. People in rich countries use around ten times as much water daily as compared to people in poorer countries. The indirect use of water in water-intensive industrial and agricultural manufacturing processes of consumer items including fruit, soybean crops, and cotton accounts for a significant portion of this. Due to the globalization of many of these production chains, a large amount of water in developing nations is utilized and polluted to manufacture items for consumption in rich countries.

Economic activity, spanning from industrialization to services like tourism and leisure, is rapidly expanding. This expansion necessitates more water services, supply and sanitation, putting further strain on water resources and the natural ecology. Because thermal power generation utilizes water for steam generation and cooling, the estimated 50% increase in global energy demand by 2040 will raise the demand for efficient water use and may transfer certain agricultural water supplies to industrial use (Lopez et al., 2009).

3.4.3. Climate Change

Because of the strong relationship between the climate and the hydrological cycle, climate change might have a huge impact on water resources around the world. Rising temperatures will enhance evaporation, which will result in more precipitation; however, there will be regional variances. Droughts and floods may become more common in different places at different periods, and mountainous locations can expect major variations in snowfall and snow melt. Higher temperatures will have unanticipated consequences for water quality. Increased eutrophication is one of the possible consequences. Changes in the climate could also lead to an increase in demand for agricultural irrigation, sprinklers, and maybe pools. The evidence is abundant now that increased hydrological variability or climate change has a tremendous impact on the hydrological cycle, water availability, water requirements and water distribution across the global, regional, Basin or community level and will continue to have a deep effect on the water sector. The FAO states that 1.9 billion people will be living in countries or regions with an absolute shortage of water by 2025, and two-thirds of the world's population may be under stress (Araujo et al., 2015). The World Bank argues that the future patterns of water supply and usage can be dramatically modified by climate change, therefore increasing the levels of water stress and insecurity in both global sectors and water-related sectors.

Figure 3.7. A global catastrophe that is developing and is reaching over 785 million people all over the world is the Clean Water Crisis. Physical water shortage is based on the need for clean water. Although clean freshwater is a critical resource for life, there is a lack of access to water for 1.1 billion people and a shortage of 2.7 billion of water every single month.

Source: https://www.wikiwand.com/en/Physical_water_scarcity.

Another common view is that climate change has reduced the volume of freshwater accessible. Climate change caused glaciers to retreat, water and river flow decreased, and lakes and pools to shrink. Many aquifers have been pumped over and are not easily recharged. Although the complete quantity of freshwater is not utilized, much has been polluted, salty, and unfit for drinking, industry and agriculture. Farmers must endeavor to enhance production to meet increasing food requirements to prevent a worldwide water disaster, while industries and cities are seeking ways to more effective use of water.

Retreating glaciers in the Himalayas can decrease the flow of water by up to two thirds in the summer. In the area of Ganges, 500 million people would need water (Leahy & Anderson, 2008). Climate change is having an impact on drinking water, which depends on the five main rivers of the Himalayan Mountain in the Hindu Kush area (HKH). While the effects will vary from place to place, the volume of the melted water is projected to increase initially because of the re-treatment of glaciers and gradually decrease as the glacier mass decreases. Climate change is making it difficult to improve access to safe drinkable water in those locations where accessible water is decreasing.

HKH experiences quick urbanization which leads to serious water shortages and water resource pressure. Rural communities would also suffer from a lack of appropriate infrastructure for water management and restricted access to potable water. The lack of drinking water will make more people migrate. That is because the poor are left behind, deaths and suicide are higher, and urban areas are more urbanized. This will increase inequality.

3.4.4. Population Growth

About 50 years ago, water was commonly perceived as an endless resource. At the time, the number of people on the earth was less than half as high. The people were no more affluent than they were today, ate fewer calories, and ate less meat. A third of the volume of water we extract from rivers was needed. Today, there is significantly more rivalry for water supplies. Their consumption of water-treated meat is growing and there is increased competition for water from industry, urbanization of bio-fuels and water-dependent food items on the planet. It is because seven billion people are on the planet. More water is needed for food production in future, as the population of the earth is projected to increase by 9 billion by the year 2050.

Figure 3.8. Many water supplies that continue to promote ecosystems and feed an increasing number of people are stressed. Waterways, lakes, and rivers, dry or become too polluted to be used. There have been more than half the wetlands on the planet. Farming utilizes more water than any other source and, through inefficacy, wastes much of it. Climate change alters global weather and water patterns, causing droughts and shortages in some areas and flood elsewhere.

Source: https://www.worldwildlife.org/threats/water-scarcity.

In 2000, there were 6.2 billion people worldwide. The United Nations projects that by 2050, an additional 3.5 billion persons will suffer from the majority of growing water stress in emerging countries (Larson & Lach, 2008). Therefore, water demand will increase unless the water conservation and recycling of this crucial resource are increased accordingly. Based on the data supplied by the United Nations, the World Bank continues to stress that access to water for food production is one of the primary issues in the next decades. The necessity of sustainable water management must be balanced with the impacts of climate change and other social and environmental factors. Water access must be regulated.

3.4.5. Rapid Urbanization

There is an accelerating tendency towards urbanization. In high-density urban areas, small private wells and septic tanks that work well in communities of low density are impossible. Urbanization needs substantial investments in water infrastructure to supply people with water and process wastewater concentrations – from the individual as well as commercial people. These contaminated waterways must be treated, otherwise pose intolerable threats to public health. Groundwater is consumed quicker than restored in 60% of European towns with more than 100,000 people. Though there is still some water accessible, it costs more and more to catch it.

3.5. IMPACTS

The water problem occurs in several main ways:

- Middle East and North Africa food security.
- Insufficient access for some 885 million people to safe drinking water.
- Insufficient access for 2.5 billion people to sanitation, often leading to water pollution.
- Over-drafting of groundwater (over consumption) leads to reduced farm output.
- Excessive water use and pollution damage biodiversity and water resources.
- Sometimes wartime leads to regional conflicts over restricted water supplies.

3.5.1. Environment

Water scarcity has numerous detrimental environmental implications, including harmful environmental consequences on lakes, rivers, swamps, wetlands and other fresh water supplies. The ensuing overuse of water, often in the irrigation fields of agriculture, causes environmental damage in several ways including increasing salinity, nutrient contamination and loss of flood-lands and wetlands. In addition, water scarcity makes the rehabilitation of urban streams difficult for flow management.

More than 50% of the wetlands of earth have been damaged and gone within the previous hundred years. Not only do they support the cultivation of rice and other food crops, as well as the protection from storms and floods, they provide habitats for diverse populations such as mammals, birds, fish, amphibians and invertebrates, they also provide water filtering. There have also been freshwater lakes such as Central Asia's Aral Sea. Once the fourth largest freshwater lake in the world, it has lost more than 58,000 square kilometers in surface area.

Another impact of water shortages is the loss or the progressive sinking of land formations. The US Geological Survey estimates that over 17,000 square miles of farmland have harmed more than 80% because of groundwater use in 45 U.S. states (Ashton & Turton, 2009). Freshwater supplies are essential to vegetation and wildlife. The marshlands, sweeps and riparian regions are more subject to a sustainable water supply, but when the available water decreases, forests and other highland ecosystems face major changes in productivity. In the case of wetlands, a large area has simply been derived from the exploitation of wildlife to feed the increasing population of humans.

However, there has been a progressive decline in freshwater inflow in other locations since upstream sources for human use are distracted. More than 80% of all wetlands were filled in seven states in the United States in the 1980s, as Congress did not lose wetlands.

The ensuing loss of biodiversity has also led to a large loss of wetlands across Europe. In Scotland, for example, many bogs have formed or decreased by the development of the human population. Portlethen Moss in Aberdeenshire is one example.

Figure 3.9. There have been tremendous gains in the accessibility of clean water, with people without access to clean water falling from 1.1 billion in 2000 to 785 million in 2017. But the benefits of clean water may continue to be multiplied by improving sanitation and hygiene.

Source: https://www.worldvision.org/clean-water-news-stories/global-water-crisis-facts.

A significant shift happened on Madagascar's highlands plateau, removing practically every densely wooded vegetation between 1970 and 2000. Approximately 10% of the country's total natural biomass has been removed by slash and burn agriculture and converted into a wasteland. These consequences were caused by overcrowding and the need to feed destitute indigenous peoples, but the negative repercussions included widespread gully erosion which in turn produced badly silted rivers, running red decades after deforestation (Altenburger et al., 2015).

This removed a great deal of fresh water and also damaged many of the riverine ecosystems of several large western rivers. Various types of fish were driven to an end, some of which are lost, for example the troubled forms of Tokyo's coral reef in the Indian Ocean. In October 2008, Nestlé President and former Chief Executive Officer Peter Brabeck-Letmathe warned against future depletion of world water supplies through the manufacturing of biofuels.

3.5.2. Water Shortages

Waterborne infections are one of the top causes of death worldwide due to a lack of sanitation and hygiene. Waterborne illnesses are a primary cause of death in children under the age of five. According to the World Bank, contaminated drinking water, inadequate sanitation, and poor hygiene are responsible for 88% of all waterborne infections. Water is the fundamental tiny balance of safe water supplies, but controllable elements such as water organization and transportation contribute to the continued shortage.

A 2006 UN report focuses on administration as the core water dilemma, which states, "Everyone has enough water" and "Water scarcity is typically due to mismanagement, corruption, lack of suitable institutions, bureaucracy and underinvestment in physical and human infrastructure." Official data also suggest that access to safe water has a clear association with per capita GDP. The main reason for this was the lack of property rights, government rules and water subsidies which caused the prices to become too low and the consumption to be too high, which led to the privatization of water.

3.6. APPROACHES

3.6.1. Cooperation

The probability of conflict is allegedly increased if the rate of change within the basin exceeds the institutional ability to absorb this change. While water emergencies can closely be related to regional tension, history has demonstrated that cooperation situations greatly outweigh acute water-related disputes. Nevertheless, in many regions of the world, especially in developing countries, regional conflicts can be caused by a lack of collaboration mainly because of the water availability, use and water management problems. The conflict over the Grand Ethiopian Renaissance Dam has intensified in 2020, for example, between Egypt and Ethiopia. Egypt considers the dam as an existential threat and worries about the water that it receives from the Nile being reduced by the dam.

Notwithstanding the mutual antagonism of both countries, the Indus River Commission and the Indus Water Treaty of 1960 have withstood two wars between India and Pakistan, and have proved a successful dispute resolution mechanism through a framework to discuss, examine, and share information. Since 1957 the Mekong Committee has been operating and has survived the 1955–1975 Vietnam War. Regional instability, on the other hand, leads to a lack of cooperation institutions in countries such as Egypt's high damon plans in the Nile (Kirschke & Newig, 2021).

Figure 3.10. In previously impacted areas, global warming is anticipated to increase the number of water strained places and exacerbate water stress. Subtropical locations such as Australia, the southern United States and countries of North Africa should get warmer and more severe droughts, although it is forecast to be more severe when precipitation occurs in these places. Climate scientists suggest that weather in tropical places also varies.

Source: https://www.cfr.org/backgrounder/water-stress-global-problem-thats-getting-worse.

However, no global organization has been supervising Tran's boundary water resources management since 2019 and international cooperation has taken place through ad-hoc cooperation among agencies such as the Mekong Committee which has been established as a result of the UNICEF-US Bureau of Reclamation Alliance. Strong international institutions are seemingly well established—encouraging early action and management and preventing costly processes for settling disputes.

Almost all resolved disagreements are commonly seen as a "need-based" paradigm instead of a "right-based" paradigm in negotiations. "Needs" are defined by irrigated land, population and project technicalities. The only water contract ever signed at the Jordan River Basin, which concentrates upon the need, not on the rights of riparians, reflects the success of a paradigm based on necessities. In Bangladesh, the water distribution on the Ganges River is determined by irrigation conditions in the Indian subcontinent (Krampe, 2017).

A needs-based regional strategy aims to satisfy people with their water requirements, to ensure that quantitative minimum requirements are met. It resolves the conflict which develops when countries see the Treaty from a point of view of national interest and move away from a zero-sum approach to a positive-sum, inclusive approach that allocates water and its benefits fairly. This means that water usage methods are both equitable and effective, in particular in the context of water shortages. In the context of sustainability, the combination of these two performance variables should make continued collaboration among all stakeholders particularly desired in the learning mode.

Figure 3.11. The United Nations recognizes that Egypt will confront a significant water shortage in the next few years, requiring more efficient management of the scant resources available. In order to avoid the existing waste, the irrigation systems should be modernized. If water shortages are not handled in responsible manners, a country of about 80 million people will be in danger of running out of water in the next decades. This may lead to a humanitarian disaster, likely destabilizing the Mediterranean area with unforeseen implications.

Source: https://www.seametrics.com/blog/global-water-crisis-facts/.

In partnership with the Swiss and Swedish Governments, a blue peace framework designed by the Strategic Foresight Group provides a unique structure of policy that supports sustainable water management and peaceful cooperation. The possibility of peace can be increased by making the most of shared water resources through cooperation instead of just being assigned to each country. In (for example) the Middle East and Nile basin, a Blue Peace method has proven to be effective. The no-limit foundation is there and charity: water provides access to clean water as leads.

3.6.2. Water Conservation and Production

Some (fresh) water protection and manufacture using diverse technologies are the solutions to the numerous national water crises.

- Dehumidification and Solar humidification

  Many air generators work much like that of a dehumidifier: air passes through a cooled spiral, condensing water. The positives sides are their low price, the lack of heavy metals as well as germs that improve the health of populations and the diversity of air as water sources without the requirement for a nearby ocean, river, and lake.

- Clean water technology

  Wastewater treatment helps conserve natural waterways and places like Singapore has started using treated wastewater as a source of potable water. Machines for desalination are meant to remove mineral elements from saline water. Desalination is more broadly the elimination from the particular compound of salts and minerals, the end to the worldwide water issue can be seen in energy efficient desalination with an electricity consumption of under 1,0 kWh per cubic meter of freshwater. Many firms, notably Siemens and TS Prototype-Creation, have developed technology to this effect. 1.0 kWh is not much higher than the need in Germany for water pumping. The 0.45 KWh cubic meter distilled water generates IBTS Greenhouse, which was conceived for wastewater treatment (Kim et al., 2015). Alongside the processing of wastewater and desalination, the number of people affected by water shortages globally might be significantly reduced. However, it is also necessary to examine the economic and environmental negative consequences of these technologies.

Figure 3.12. The pressures on urban water resources are predicted to increase as cities and towns grow – according to the analysis, by 2030 demand would double the supplies available. In Indian gross domestic products, water scarcity will also cause for 6% loss (GDP). However, some Indian states do better management work than others. Gujarat topped the rankings in the west of the report. Madhya Pradesh in Central India and Andhra Pradesh in South.

Source: https://www.bbc.com/news/world-asia-india-44492994.amp.

3.7. ENGINEERING TECHNOLOGIES

With classical engineering, a number of water shortages can be tackled, typically with immediate advantages. Infrastructure repair is one of the most evident answers. Challenges in the repair and building of infrastructure can be identified to reduce installation and maintenance costs, in particular in less developed nations As well as to build engineering solutions that help the environment and tackle climatic changes.

With approximately 70% of all freshwater resources for farming, improvement of irrigation technologies is another key answer (Jackson, 2006). As the principal means of irrigation, many agricultural regions depend on simple floods or irrigation of the surface. However, floods frequently accumulate fields with more water than crops, and large volumes of water are lost for evaporation or transit from their spring. Educating farmers about possible loss of water from these methods can assist minimize wastewater consumption in agriculture by setting clear reduction targets and funding irrigation upgrades as well as water preservation technologies.

Desalination was suggested in locations that had access to brackish soil or water to alleviate water shortages problems. Indeed, wastewater in certain densely populated dry areas, such as Saudi Arabia, already is the major source of municipal water supply. The desalination technology currently in operation, however, takes considerable energy, usually in the form of fossil fuels, therefore the process is costly. This is why it is normally only employed in cases with inexpensive sources of fresh water. Moreover, there are major environmental issues in terms of emissions of greenhouse gases and brine wastewater generated by desalination plants.

Wastewater can be a significant resource in urban or urban areas where there is a growing population and limited water supply. As well as decreasing the burden on limited supplies of freshwater, wastewater reuse can improve the health of rivers and lakes by lowering their contaminated effluent dumps. Wastewater may be recovered and reused for irrigation, reclaiming and restoration of crops and landscapes. It is technically conceivable to reclaim drinking or household use, but reuse is faced with considerable societal resistance. In urban areas globally, the creation of water recycling plants is more and more frequent. Wastewater is suggested as a technique of efficiently growing such water-intensive crops while fostering renewable sources of electricity.

Figure 3.13. The lack of access to clean water adds more pain to poor wastewater treatment systems. Almost 80% of wastewater is disposed of in the environment, and only approximately 20% of wastewater has been cleaned up to a certain level based on the current data on wastewater collection and treatment in metropolitan India.

Source: https://www.wgsol.com/safe-and-clean-water-scarcity/.

Harvesting rainwater can greatly reduce demand for public food supply and strain on the storm water infrastructure for non-potable uses such as gardening and washing clothes (Hirji & Ibrekk, 2001). In big towns and municipalities with a high water stress, including Mexico City, there are considerable savings in demand and supply of drinking fresh water. In many locations rain and other rainfall systems are encouraged and subsidized. However, in other locations, especially in the West of the United States, gathering rainwater is considered a matter of water rights, and such collections are restricted. Catchment systems collecting and percolate runoff into the ground are also effective for groundwater recharging.

CHAPTER 4

MANAGEMENT OF WATER RESOURCES

CONTENTS

Management of water resources is the planning, development, distribution and management activity for the best use of water resources. It is a part of the management of the water cycle.

Water cycle management is a holistic method for all water cycle planning, development, operational and strategic decisions. Most significantly the management of the water cycle guarantees the supply of clean water to be used and the safe disposal of treated water.

Water is in a continuous cycle in an unstirred setting and may usually be used for the majority of nature as in each step of the cycle. The natural cycle is interrupted following human contact.

Runoffs accumulate debris, and these debris and other contaminants in urban agricultural regions cannot be cleaned of water by the natural purification process. Moreover "used water" can if not adequately managed, be exceedingly hazardous to nature from houses and businesses.

Figure 4.1. The global water cycle.

Source: https://en.wikipedia.org/wiki/Water_cycle_management.

Water cycle management serves to meet human and environmental goals in various disciplines of environmental science and engineering. Typically, the management of the water cycle can be separated into six sub-flows that deal with the problem from different perspectives: meteorology, hydrology, the management of water resources, water technology, water conservation and environmental monitoring.

The political and socio-economic aspects of the management of water cycles are also recently taken into account because of the unequal distribution of freshwater quality and quantity globally.

For human survival, water is necessary. The management of water resources must change to address the present and future water distribution problems. Decision-making will be considerably more complex with the increasing ambiguities of global change and the long-term repercussions of management activities. Continued climate change is likely to result in unpleasant circumstances (Ghosh, 2017).

As a consequence, innovative management solutions are being developed to avoid retrograde water resource distribution.

Preferably, water resource management plans take account of all competitive water requests and aim to equitably distribute water to meet all uses and requirements. In actuality, this is seldom practicable, as with other resource management.

The viability of present and future water resource allocations is one of the main challenges for our water resources in the future. The significance of water management is growing considerably as water scarcity is increasing, establishing equilibrium between the requirements of humans and the crucial step in water sustainability.

Water technology is an important field aimed at ensuring clean water and water security and can be used in any phase of the water cycle. In addition, water technology can be separated into sub-sets of architectural water technology, water treatment and wastewater treatment.

In structural water engineering, structures that regulate the water resources are built, repaired and maintained. Reservoirs, dams, sewage stations and water systems are the most essential ones concerning water cycle management. These are all key characteristics of water's natural process.

Re-use therapies are more significant than static infrastructure such as dams when it comes to water cycle management. Any method used for removing pollutants from water and improving water quality is water treatment.

Figure 4.2. Gordon Dam, Tasmania, Australia. Dams are a part of static structural water engineering.

Source: https://en.wikipedia.org/wiki/Water_cycle_management/

Water can be assigned for residential use such as drinking water, provision for industrial or agricultural use and the manner of treatment relies on the function of final use. Water treatment can also be used to safely bring water back into the environment. Functionally, sewage treatment is quite equivalent to water handling, but it does handle wastewater influenced by human use: residential as well as industrial effluent wastewater. The purpose of the treatment of wastewater is to purify wastewater and provide water for recycling. Water is treated using a variety of techniques such as chemical treatment, bacterial usage, biological processes and ways of UV disinfection. Significant levels of noxious chemicals like drugs are still discovered in the return environment and water cycle notwithstanding elaborate treatment procedures.

A hydrosphere that is sustainable is required by the growing population. Fresh water is a requirement that must be met, both in the now and in the future. Climate change is a major component in this growing demand. By using policy management to secure water supply for generations to come, reducing energy consumption and protecting local fauna and migratory birds, freshwater habitats, and ensuring their inhabitants water quality.

The major activities in the preservation of waters include reducing water loss, water usage and waste, preventing the decrease in water quality and enhancing water management methods that reduce water consumption.

Environmental monitoring refers to the processes and actions to define and monitor environmental protection. In the production of environmental impact assessments, environmental monitoring can be used, and in many cases where human activities provide a danger of detrimental impacts on the natural environment. The objectives and explanations for all monitoring techniques and programs frequently serve to establish the existing state of the environment or to set trends in environmental conditions. In all situations, surveillance results are evaluated, examined and reported statistically. Thus, before commencing actual monitoring, the design of a monitoring program should consider the ultimate use of the findings.

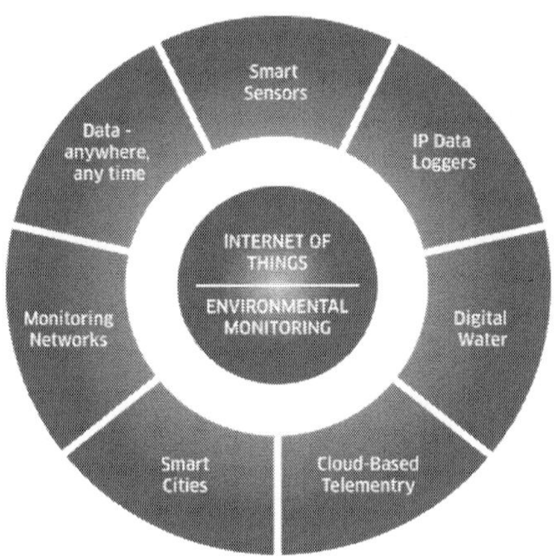

Figure 4.3. Environmental Monitoring.

Source: https://www.omnisci.com/technical-glossary/environmental-monitoring.

Agriculture is the world's main water consumption activity, absorbing 70% of the world's freshwater. With the world's increased population consuming more food (now above 6%), the industry and urban expansion are growing, and rising biofuels also require a share of freshwater resources, result in an essential water shortage. Water shortages are becoming a significant problem. The International Water Management Institute of Sri Lanka evaluated water resource management in agriculture in 2007, to see whether the world had enough water to feed its expanding population or not. It examined the existing global amount of water available for agriculture and identified water scarcity sites. A fifth of the globe's citizens, over 1.2 billion, have found that they live in areas of physical water abuse, with insufficient water to satisfy all their requirements. An additional 1.6 billion people live in regions with a shortage of economic water, which prevent authorities from satisfying their demand for water because of their lack of investment in water or inadequate human capability (Grizzetti et al., 2016).

The analysis deemed it plausible to produce enough food that would be necessary in future, but the ongoing food production and environmental patterns in many regions of the world may result in catastrophes. With regard to food production, the World Bank is focusing on agricultural food production and water resources management to promote a significant and expanding debate as an increasingly global issue. The authors of the book, Out of Water: From Abundance to Scarcity and How to Solve Water Problems, set out a six-point strategy to address water challenges around the world are 1) improving water data; 2) environmental conservation; 3) water management reform; 4) reinvigorating farm water use; 5) managing urban as well as industrial demand; and 6) empowering the poor and women in the management of water resources. Farmers will need to make efforts to improve production to meet expanding food demands to prevent a worldwide water disaster, while industries and cities can use water more efficiently.

Figure 4.4. Out of Water: From Abundance to Scarcity and How to Solve the World's Water Problems

Source: https://www.goodreads.com/book/show/9250696-out-of-water.

FAO and the World Water Council (2015) have assessed that food production will continue to be enough to support a world population of $9–10 billion by 2050 with proper investment and policy initiatives; however, food and food insecurity remain unchanged in many places. Food safety policies that expressly take into account a structural food supply shortfall and trade agreements that protect against food volatility are increasingly needed by countries in water-based regions (Geissen et al., 2015). Because huge sectors of the population in developing countries will strive to enhance their quality of living, increased strains on the water for agricultural productivity may occur. When compared to bread, eating more beef necessitates a lot more water.

Figure 4.5. FAO and the World Water Council (2015)

Source: https://www.worldwatercouncil.org/zh-hant/node/1561.

The earth's carrying capacity is expanding dramatically as a result of technological advancements and urbanization, both of which are fuelled by increased economic opportunities. Rapid urbanization is happening all around the world, but it is especially noticeable in emerging economies and underdeveloped countries. China and India, which are home to many of the world's megacities (cities or urban regions with populations over 10 million), are both growing at breakneck speed. The quantity of megacities is expected to rise further, reaching around 50 by 2025 (Gao et al., 2014). Water scarcity is a very common and widespread problem in emerging economies. The eastern hemisphere's freshwater resources are depleting at even a quicker rate than the poles. Today, millions of people lack access to safe drinking water. Contaminated freshwater resources, depleted groundwater resources, poor harvesting abilities in rural regions, poorly designed and managed water supply systems, high levels of unofficial water usage, and inadequate technological and water management capabilities are all contributing factors.

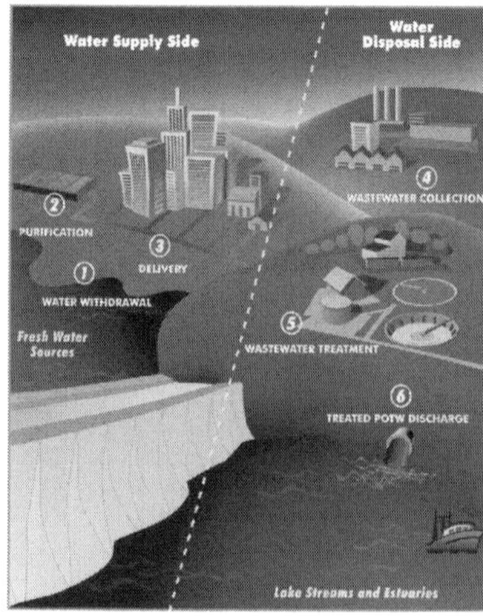

Figure 4.6. Typical urban water cycle in the United States (POTW = publicly owned treatment works; a municipal sewage treatment plant).

Source: http://css.umich.edu/factsheets/us-wastewater-treatment-factsheet.

Agriculture contends with industry and municipal consumers for clean water resources in the outlying areas of urban centres. Conventional water sources are getting polluted with urban runoff as a result of this rivalry. Farmers frequently have little choice but to irrigate their crops with dirty water because urban areas offer the best prospects for selling goods. Based on how well a city's wastewater treatment system is established, drinking this water can pose serious health risks. Pollutants can be found in wastewater from cities. Excessive nutrients, salts, and a wide spectrum of pathogens are commonly found in wastewater from kitchens, toilets, and storm water runoff. Heavy metals as well as residues of antibiotics and endocrine disruptors like oestrogens, may be found in wastewater.

Developing nations have the lowest rate of wastewater treatment. While, creative public-private partnerships have raised wastewater treatment to greater extend than local recycling capacity in several developing country desert cities Sometimes, bacteria from sewage pollute the water that farmers use to irrigate their crops. Bacteria, viruses, and parasitic organisms are the microorganisms that offer the greatest harm. These viruses have a direct impact on farmers' health, as well as consumers of contaminated products. Diarrhoea is a frequent ailment that kills 1.1 million people each year and is the second leading cause of newborn death (Ali et al., 2019). Many cholera epidemics are linked to the usage of sewage that has not been properly treated. As a result, initiatives to prevent freshwater contamination play a critical part in the global health fight.

Using a concept known as the "multiple-barrier strategy," scientists have been striving to identify ways to decrease food contamination. This entails examining the entire food production chain, from planting crops to purchasing them in markets and then eating them. Then think about where one might be able to establish a contamination screen. Improving irrigation practices, boosting on-farm wastewater treatment, removing pathogens, and adequately cleaning crops after harvesting in markets and restaurants are all barriers to overcome.

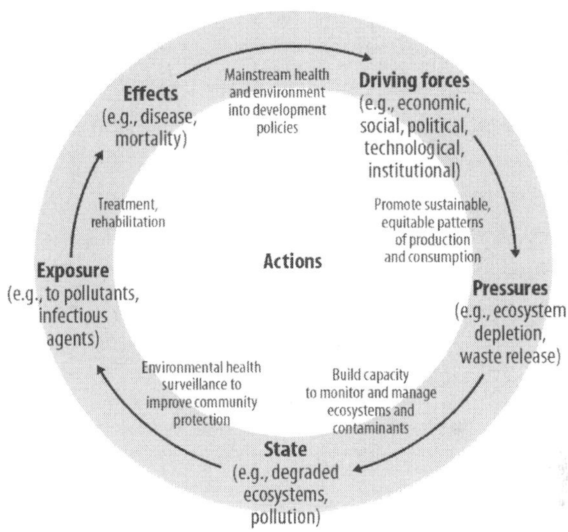

Figure 4.7. Components of multiple-barrier strategy.

Source: https://www.researchgate.net/publication/262912150_Ecohealth_and_Watersheds_Ecosystem_Approaches_to_Re-integrate_Water_Resources_Management_with_Health_and_Well-being/figures?lo=1&utm_source=google&utm_medium=organic.

4.1. CONTEXT AND CHALLENGES IN WATER MANAGEMENT

Most nations are currently putting enormous strain on their water supplies. The world's population is rapidly increasing, and projections predict that, if existing growth rate continue, the globe would confront a 40% gap between expected demand and available water supply by 2030. Chronic water shortage, hydrological unpredictability, and extreme weather events (floods and droughts) are also seen as major challenges to global security and prosperity The importance of water shortages and drought in exacerbating fragility and conflict is becoming more well recognized (Fuhrmann, 2012).

By 2050, feeding 9 billion people will necessitate a 60% rise in agricultural production (which currently utilizes 70% of the resource) and a 15% rise in water withdrawals. Apart from the rising demand, the material is already in short supply in many regions of the world. According to estimates, 40% of the world's population lives in water-scarce areas, and this dilemma affects about 14% of global GDP. By 2025, around 1.8 billion people will be living in water-scarce regions or countries. For so many nations today, water security is critical – and almost always growing – issue.

Climate change will exacerbate the problem by disrupting hydrological cycles, leaving water more uncertain, and escalating the regularity and severity of floods and droughts. The 1 billion people who live in monsoon basins and the 500 million people who live in deltas are particularly vulnerable. Droughts, for example, impose constraints on the impoverished and those living in rural areas, which are strongly reliant on precipitation patterns for sustenance. Flood losses are projected to be $120 billion per year (mostly from the destruction of property).

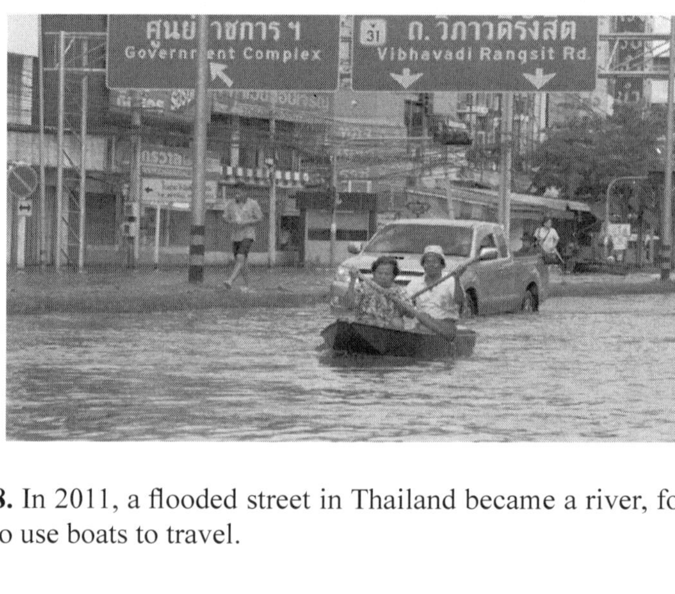

Figure 4.8. In 2011, a flooded street in Thailand became a river, forcing some residents to use boats to travel.

Source: https://thecityfix.com/blog/number-people-affected-floods-will-double-2030-samantha-kuzma-tianyi-luo/.

Water security is further hampered by the fragmentation of this resource. About 276 transboundary basins are shared by 148 countries, accounting for 60% of world freshwater flow. Likewise, 300 aquifer systems are transnational, implying that 2 billion people rely on groundwater around the planet (Figuères et al., 2003). Fragmentation concerns are frequently mirrored at the national level, necessitating collaboration to provide effective water resource management and development solutions for all riparian's. Governments will need to enhance the way they regulate their water resources and related components to deal with these complex and interconnected water concerns.

Users will have to engage in capacity building, data management and (natural and human) infrastructure projects to enhance the security of water under the context of increased demand, water shortages, increasing unpredictability, extreme conditions and fragmentation obstacles. To better allocate, control, and preserve water resources, systemic mechanisms such as legal and regulatory structures, water pricing, and tariffs are required.

Resource monitoring, decision-making under ambiguity, systems assessments, and hydro-meteorological forecasting and warning all require information systems. In addition to searching for alternatives for improved storage and recovery, exploring innovative solutions to improve productivity, conservation and protection of resources, storm water reclamation and wastewater and the development of non-conventional water sources. Such developments will be important to increasing worldwide water safety, ensuring quick transmission and the right modification or application.

4.2. STRATEGIES FOR WATER MANAGEMENT – WORLD BANK

The World Bank aims to help nations fulfil their economic growth and poverty reduction targets based on the Sustainable Development Goals (SDGs). SDG 6.5 addresses water management in particular, although for its accomplishment other SDGs and objectives need the management of water resources. The Bank has therefore a strong interest in supporting nations to achieve water safety through good and robust management of water resources.

Figure 4.9. Sustainable Development Goals (SDGs).

Source: https://www.un.org/development/desa/disabilities/envision2030.html.

The aim of water management is water security. In the context of increasing climatic and non-climate uncertainty, it is difficult but not impossible to "predict and design" a single path to water security for a rapidly growing and urbanizing world population. According to the World Banks assessment, we need capacity-building, flexibility and endurance for the prospective planning and management of water resources to bolster water security.

Water Resource Management (WRM) is the procedure through which water resources, both in respect to water quantity and quality are planned, developed and managed in all water uses. It covers institutions, infrastructure, incentives and water management data networks (Fang et al., 2013). The management of water resources aims to exploit the advantages of water through adequate water supplies, food production, power production, internal water transhipment, leisure water services, the maintenance of healthy water-dependent ecosystems and protection of the aesthetic and spiritual values of lakes, rivers, and estuaries. The management of water resources also involves control of the risks associated with water, including floods, drought and contamination. The intricacy of water/household, economic and ecosystem linkages necessitates integrated management, responsible for the synergies and compromise between the greatest admixture of water diverse uses and values. Water safety is realized by leveraging the productive potential of water and managing its destructive potential. Water safety is different from food security or energy security notions because it is not only difficult to ensuring enough resource supplies, but also reducing the dangers posed by water when it is not handled effectively. Water safety represents activities that can or can be taken to guarantee that water resources are used sustainably, to provide reliable water utilities and to manage and minimize the risk associated with them. A dynamic structure that extends beyond single-issue objectives such as water shortages, pollution or access to water and sanitation proposes more generally considering societies desires decisions and accomplishments in water management. It is a progressive political objective that transforms the values of societies and the economies and affects the exposure to water-related hazards and sensitivity of societies. It has to deal with equity difficulties.

The World Solutions Group (WSG) supports the bank's research, consultative and operational commitments to help customers reach their water safety objectives. A better integrated and more long-term concept of water management will need to achieve water safety in the setting of increasing water shortages, improved uncertainty, degraded water quality and aquatic ecosystems and increased droughts and flooding. The main themes will be to ensure water resources sustainability, to enhance climatic resilience and to strengthen integrated management to accomplish Global Practice Goals (GP) and SDGs. The GSG works directly via water management projects or multi-sector projects with many GPs and cross-cutting solutions fields (CCSAs), partly through farming, energy, environment, climate or urban initiatives.

Figure 4.10. The World Solutions Group (WSG).

Source: https://wsg.ag/john-bekiou.html.

4.3. RESULTS OF MANAGEMENT EFFORTS BY THE WORLD BANK

Robust responses for complicated water issues include state-of-the-art expertise and technology that are incorporated into water projects to enhance their effectiveness. New insights derived from the worldwide experience of the World Bank Group and partner skills cover global shortages in knowledge and change the design of water investment programs to produce results. Over several years, programme commitments are aimed to improve the economic situation and to better the lives of millions of poorest people around the world in a significant manner over the long run (El Bedawy, 2014).

An instrument for analysing the state and trends of water resources, water services and water threats, including climate change, transboundary waters and virtual trafficking, is the Water Security Diagnosis Tool. The tool will assist countries to assess if or to what degree water-related variables affect the population, the economy and the environment and to identify if or to what degree water-related issues present developmental and well-being opportunities.

By adjusting its businesses to meet today's needs, the World Bank is working actively to confront emerging globalizing problems.

- The Bank has promised to double its support for fragile countries through the International Development Association (IDA) 18. The Water Global Practice efforts on weakness, conflict and violence are being led by a worldwide guideline report. A report titled Water Security in Fragile Contexts, Turbulent Waters, illustrates what transpires when organizations in vulnerable nations do not address the wider spectrum of water-related concerns.
- Work spanning domains ensures that energy, environment, agriculture, urban and rural development, as well as the new global concerns, are resolved to water issues.
- The Thirsty Energy Initiative, for instance, addresses the growing difficulties of water and energy worldwide through a stronger integration among countries in water and energy management. In South Africa, for instance, the bank works with partners to include economic data regarding water into energy-optimization systems, a nation with complex water challenges and big energy development plans. Thirsty Energy helps to integrate anticipated water limitations into the country's energy plan in China as well.
- The bank is financing the Mekong Delta Global Environment Fund Integrated Climate Resilience and Sustainable Livelihoods Project, which is working for sustainable development in Vietnam. The Mekong Delta is among the most susceptible to the effects of global warming by increasing investments in delta as well as upstream activities in the Mekong Basin, along with climate change (Badham et al., 2019). To enhance the area's

climate change resilience, there is a growing demand for increased research and development to move from traditional conventional lifestyles to more environmentally resilient practices and help to mitigate climate change. The purpose of this project is to increase the research and innovation capability of research organizations and institutions in the designated regions in the Mekong delta of Vietnam to create and implement climate robust and climate-friendly natural resources management methods.

The bank also promotes restructuring commitments and activities aimed at optimizing water and other infrastructure physical, green and co-utilization advantages.

- For instance, the Mozambique Water Development Project blends multifunctional water use, and management, to boost the Corumana Dam's advantages.
- The bank will create a knowledge platform to stand on the NBI concept and support bank interactions with nature in emergency risk management, the environment and water management projects to enhance awareness of natural-based infrastructure (NBI).

• The institutional and political factors include a considerable number of World Bank-supported water resource management initiatives.
  - The IWRM project in Peru is intended to develop focused water resources management capability at the national level and 10 selected water river basins in Peru, to plan, control, and manage water resources (Bagatin et al., 2014).
  - The knowledge item "Water and Learning approaches" addresses reassignment concepts experimentally.

• Support for cooperative transboundary water management can contribute significantly to the improvement of efficient and fair management of water resources through 263 transnational rivers around the world. The Bank helps cross-border waters by providing the knowledge and loan portfolio of Multi-Donor Trust Funds (MDTF)
  - The MDTF, run by the World Bank and funded by the European Commission, the Swiss Economic Affairs Secretariat, USAID, DFID, is the Central Asian Energy-Water Development Program (CAEWDP). The MDTF enhances energy and water sustainability by taking use of expanded Central Asian cooperation, comprising all five Central Asian countries and Afghanistan.

- The CIWA is a World Bank MDTF funded by the European Commission, the Netherlands, Norway, Sweden and the United Kingdom and overseen by the World Bank. The trust funds upstream operations in African International Rivers, of which 75% are allocated to four priority areas – Nile, Niger, Volta, and Zambezi.
- The MDTF, run by the World Bank and funded by the UK, Australia and Norwegian governments of South Asia, is the South Asia Water Initiative. In the main Himalayan River systems—Indus, Ganges, and Brahmaputra—the Trust Fund gives funds to recipients.
- In the Mekong River Basin, the Bank is enhancing the unified water resources management and disaster response and management skills of the River States such as Cambodia, the Democratic People's Republic of Lao, and Vietnam by directly working with the Basin-wide Mekong Commission.
- The bank also invests in technologies such as ROTE (Retooling Operations with Transboundary Impacts) to discover mechanisms to encourage riparian country collaboration to mitigate transboundary damage as well as to reap advantages from investment in transboundary reservoirs.

• Including good working early-warning systems, facilities and institutional frameworks for decisive effort to handle rising unpredictability and variations in runoff and flood trends, the bank is following an integrated flood control plan.

• In Madagascar, the bank invests through strengthening organizational delivery of services and flood resilience in neighbourhoods of Greater Antananarivo, as well as in strengthening the state's capability for integrated urban management and efficient reaction to eligible disasters and emergencies.
  - The bank enhances flood protection in Argentina and increases the capacities of the agencies accountable for the coordinated surveillance and management of water resources in the Salado River Basin (Dobriyal et al., 2012).
  - Since 2007 the bank extended its reach in Poland and has helped develop robust water and information systems for the Odra and Vistula Basin to protect themselves from floods.

• The scarcity of water is also addressed in the following:

• The Water Scarce Cities Initiative originally aiming to support integrated methods in the water resource management and service provision in water-limited towns as the foundation of water safety and environmental sustainability. The project is initially focused on the Middle East and North Africa (MENA) countries.

- The small Island States- In small island republics, the problems and innovations of water management can be extremely dramatic. These countries are particularly concerned not only because they are often overlooked but also because they provide the chance to concentrate on intense reuse and development of non-conventional water resources, which are increasingly crucial for megacities and severely scarce environments. The state-of-the-art and portfolio of the bank are offered as overall research.

The World Bank also places focus on sustainable management of groundwater, which is vital to water safety in many nations.

Recognizing that groundwater is being drained more quickly than it is being filled in many regions, over years of discussions, the World Bank has cooperated with major global stakeholders in developing a framework for groundwater management. The 2030 Vision and Global Action Framework are a bold demand that governments and the global communities together take appropriate measures to secure the sustainable management of groundwater.

The main difficulty for water safety is the ability to decide enough to meet uncertainties and the requirements of the future. This is especially crucial in water projects involving long-lasting infrastructure expenditures which will benefit many future generations.

Sophisticated analytical work like the new research on Climate Uncertainty in the planning and development of water resources supports world development specialists and country customers in addressing climatic hazards. In order to ensure that climate change assumptions are taken into consideration, this judgment framework has already been used for the Arun Dam in Nepal.

4.4. RESEARCH NEEDS AND DIRECTIONS FOR WATER MANAGEMENT

In today's modern world where people can regulate the environment on our planet, fragmented reflexes and reactions to undesirable interruptions in support systems are insufficient. Our greatest challenge in this century is to sustainably provide healthy and meaningful livelihoods for all people. In order to meet this challenge, modifications will be needed in supply and beneficial use of water, food, energy and other products and services. It will involve modification of process in which items are produced and by-products are recycled and disposed of. Changes in consumer behaviour, particularly our wealthiest, will be necessary. In a nutshell, it means that we, as a society, will all need to define the technological, economic, social and political policies that will define the road towards achieving a desired, sustainable and better future through study, development, engineering and science and governance. We have considerably more efficient and effective know-how, technology and economic resources for the management of our water resources than in the past. Scientific inquiry is continually expanding into our knowledge and instruments by systematically examining the structure and behaviour of the physical and natural world.

We discover more about environmental conservation and their water needs through research. We understand that adjustments in our lifestyle and diets can also significantly affect our intake of water. We know that, before reaching our dinner table, we can limit the waste of water to create food that is thrown into various phases of the global supply chain. We know how to utilize less energy and therefore less water to produce this energy. We've got possibilities. We must make decisions. We must take decisions. Continued study is essential to assist us find and inform us about these ever-changing choices. Unlike some other areas, such as medicine and physics, in the domain of study which have no immediate advantages for years, most water resources and environmental research are projected to produce products on the market over a time period of 5 to 10 years (or less). Most of them did not encourage more innovative research aiming at long term answers, products, or techniques, as they should be, in the area of water resources and the environment (Bouwer, 2000).

4.5. THE PLANET REACTS—WILL PEOPLE?

Companies react with mitigation and adjustment actions. Some are slow, and some are swift; some take us one way, while others take us another. One of the most striking changes in civilization—global warming—and the tight ties to water—have not yet led to a coordinated effort to address risks. The result, in terms of climate and water changes, is likely to have a degrading or damaging effect on agricultural systems, natural habitats, and economic systems, in addition to the hydrological cycle itself. Scientific arguments about the seriousness of the consequences of increased greenhouse gas emissions expressed at international climate meetings are only marginally and slowly influencing political decisions and concrete action.

The scientific results are unpleasant not only for policymakers but also for current economic structures and even segments of the people at large. Opinion surveys suggest a universal fear in countries, e.g., the member nations of the Organization for Economic Co-operation and Development (OECD), about global warming and its anticipated repercussions. But our preparedness to adjust practice that is destructive to our environment remains a huge challenge internationally (Bouwer, 2002). Changing our ways towards a more desired future involve new thought and social research that leads to new ways of water management and, above all, political and social approval. The company tends to adhere to customary thought and procedures. Past investments and education tend to maintain people's perceptions, leading to policymaking and policies that are even lower than recognized solutions. Funds of physical capital are immense investments in water infrastructure (for example dams), storage and transport equipment. Naturally, for many people, the previously existing structures, and related organizations and expertise are highly significant.

However, it is fair to explore, given the increasing more widespread quality limits and supply of water, whether different land and water management policies could produce desired lives with a lower environmental risk. Road reliance is linked to the creation of policies. Some practices are "privileged," while others are overlooked. Projects and concepts in the previous category receive considerable favourable attention, and "more motivation than comprehension" is occasionally addressed. It is reasonable to see that both the populace and policy-makers are intrigued by major projects such as Hoover, Aswan and the Three Gorges Dam in the historic hydraulic civilizations and other similar projects in contemporary times. In contrast, rains fed systems are less showy and political and often less enthusiastic about budget allocations.

Figure 4.11. Three Gorges Dam.

Source: https://en.wikipedia.org/wiki/Three_Gorges_Dam.

Late John Briscoe said that every country which has effectively raised its populace out of poverty has mostly done this by increasing its basic production capability for economic growth and job creation, particularly in agriculture, energy, transportation and water. Without any such investment, which was a catapult for private-sector employment creation, no currently rich country has grown.

Every rich country has exploited more than 70% of its economically feasible hydro-electric potential by taking only one indicator. 3% of Africa's potential has been developed (Butt & Bilal, 2011). Not only was it the road which all currently wealthy nations have travelled,

but it is also the road travelled by countries which in recent decades, like China, India and Brazil, have taken their citizens out of poverty Water storage is vital and can be done from wetlands to big dams through a wide range of architecture. The World Commission for the Large Dams presented helpful guide to prevent these mistakes in the future, although errors could have occurred in the past, especially on the environmental and socioeconomic consequences of major dams.

4.6. MOVING TOWARD WATER SECURE PLANET

There is no way for a future, say, in 2050, to be ideal. A lot of "desirable" future paths exist. But one of those worlds that must not cope with water shortage is difficult to envision. In rising countries like China and India, which together represent approximately a third of the global population by 2050 (Brack et al., 2015).

The survival of healthy ecosystems will certainly be a component of any "good" future. Water is needed for living environments. Other water requirements were often seen as contrary to the need for water to sustain ecosystems.

In addition to the requirement for freshwater for ecosystem preservation, the disposal of residential and industrial wastewater sometimes without consideration for its effect on ecosystems must be eliminated.

In summary, we will need to determine how ecosystems that are fundamental to life are deprived of water and how they become poisoned with garbage affects human life and development. To tell us this, we do not need further investigation. More research on how to do this successfully and economically can profit us.

In order to make sure enough water is available for the feeding of a growing and prosperous population, supporting vital livelihood systems and producing and distributing enough other products and services, more crops are needed, more employment per drop, a better environment per drop, and improved nutrition per dropping, among many of these goals. Current trends mean that, unless we work together to overcome that increase, we may move towards more drops than less per profession, per given environment, or at a specific level of nourishment. It is more challenging to meet unfulfilled and developing human requirements and needs. Another key problem is to consider this while maintaining the operation of water (fresh and brackish/marine) and land ecosystems. The identification and assessment of strategies to solve those difficulties is certainly a topic for a big multidisciplinary joint research effort.

We need to be inventive in order to avoid a serious water catastrophe. If we keep on following a corporation approach it could cause our existing predictive patterns to fail to operate. The administrative and regulatory mechanisms that can drive us along a route to long-term environmental sustainability have to be identified, developed and then set in action (Cook & Spray, 2012). We will also need to investigate how effectively to confront financial, social and environmental concerns at the same time. The creation of conditions for human creativity to thrive in political, scientific, cultural and ethical fields has to be our primary focus. Local, regional and international leaders must put water issues on the agenda before being driven into conditions that they can no longer control. We need to find and implement imaginative long-term water management strategies that can risk mankind's future

The gaps between both the richest and poorest have been, and still are, large and growing. The access of roughly a billion people to safe water and enough food remains a dream. There is a lack of safe sanitation for more than 2.5 billion. Over the past 15 years, food insecurity has worsened as regards undernourishment, but global aggregates of per capita agricultural production and availability have improved steadily. Rising food insecurity over the last 15 years has been observed even though output and quantity have increased more quickly than the population has increased over the same timeframe (Carr et al., 2012). Productivity and supply—what is on the market after attrition and conversions—are at a level beyond the food security needs.

Figure 4.12. New report shows a slight improvement in food insecurity.

Source: https://www.world-grain.com/articles/11873-new-report-shows-slight-improvement-in-food-insecurity.

Over-feeding today is more prevalent than under-nutrition. It is estimated at 1.5 billion people aged 20 and over are over fed. In addition, between one third and one half of the farm product is lost or squandered before it reaches any dish.

The efficiency to use resources—"more crop per drop"—is crucial but the net outcome of production improvements is decreased if a substantial proportion of "crop" is not used in advantageous terms.

Not just the consumption of water and food do these kinds of faults apply. Other resources and commodities may be subject to similar observations. These kinds of defects harm us, as well as the environment and natural resources of the world, through the close relationship between food, water, energy, the environment and human well-being. In an age of increasing water shortages, the use of our technical capabilities and human inventiveness is needed to eliminate these inefficiencies in the production and supply chains. It is necessary to reduce losses and waste from things that are grown or produced by us from water. It wasn't seriously tested till now.

4.7. TECHNOLOGY

Many feel that technology is employed to attain the future that we want tools and procedures employed in the production of goods and services. Indeed, there are a large number of technology trends and developments that can improve the water industry quickly and effectively:

- Instant cybernetics, artificial intelligence and IT (smarter internet).
- Nanotechnology.
- Cost-efficient technique for energy (solar, space-based energy, algae as fuel).
- Biotechnology for the food and safety of endangered species (genetic engineering).
- Instant input on space-based monitoring systems for the environment even for remote regions of the world.
- Geo-technology to reverse global warming (e.g., giant reflectors in orbit; greening deserts; iron fertilization of the sea; aerosols in the stratosphere).
- Efficient, dependable weather and climate forecast.
- Fully substitute for fossil fuel – low-carbon companies.
- Desalinisation (with inexpensive fusion energy) becomes more cost-effective and water supply for a most large cities and urban coastal areas.
- Enhanced wastewater treatment, sanitation and recycling technologies.
- Biotech pest control technologies for enhanced farm yields.
- Engineering ecological conservation, reversal extinction and fight against invading species.
- Maps of sustainable extraction levels of groundwater resources.

The computer-based models for optimizing and simulating interactive graphics will help us continue to find the plans, designs and policies which optimize the desired effects and reduce unwanted effects and clarify the compromise between the two.

Strategies of science, technology and innovation are key components of the strategies of sustainable development. Many technologies are highly risky with uncertain returns for sustainable water management. Public-private partnership-based government funding and innovation policies can be created and executed to reduce risks and promote research, development, and technology dissemination and transfer (Cabrera & Custodio, 2014).

4.8. MONITORING

The worldwide water surveillance system has to be developed to provide the necessary information for water management and to monitor development in line with aims. This should cover a broad range of local and national and worldwide information (e.g., data tracking, public papers, and comprehensive national plans, accessible and relevant technology). A means of enabling people and organizations to access new ideas and innovations is timely, comprehensive, and future-oriented information in adequate quantity and the gradual increase in the ability to stream information into decision making. This could assist to develop a more interconnected and informed society, enabling transparency and confidence in the pursuit of joint objectives.

The data might be continuously passing across satellite links to data processing and modelling centres through, for example, a network of hand-held, cell phone sensor networks devices for climate, soil moisture and chemistry, water quality, stream flow etc. These centres might also monitor developments in storms and frequently update existing downwind or downstream region's forecasts. Such a source of information might offer any rural village the knowledge that farmers need.

For ecosystem management, an additional example is a need for far improved information and monitoring approaches and related capability. The conventional focus of ecosystem information and surveillance has been on the impacts of water consumption and interests need to go further towards ecosystem surveillance services which provide a shorter connection between ecosystems and human well-being (Zhao et al., 2017).

However, there are more than improved sensors and satellites. Significant advances in ground-based monitoring and knowledge integrating from all sources must be developed to improve the management of water supplies, including complementary aircraft monitoring systems.

4.9. URBAN DECISION SUPPORT SYSTEM (UDSS)

Urban Decision Support System (UDSS) is an urban-water system based on data that uses water-based sensors to retrieve data on water consumption in urban residential areas. With an investment of EUR 2.46 million from the European Commission, the system was created to enhance home water usage behaviour. The UDSS Smart Customer mobile device collects and sends information on devices and facilities such as dishwashers, showers, washing machines, taps to a UDSS App. The UDSS may then analyse and show homeowners which devices use the most water, which complements or behaviours to reduce water use.

This helps people more economically to manage their consumption. The UDSS, coordinated by Dr. Lili Yang, is based in the discipline of management science at the University of Loughborough, specifically, the home water benchmarking decision support system,

4.10. WATER CONSERVATION AT EPA

In the past 50 years, our need for water has increased. The population has doubled. The need to save water is crucial in at least 40 countries anticipating water shortages by 2024. In all its installations, EPA aims to integrate best practices in water management (Afroz et al., 2014).

Two main categories of equipment are occupied by EPA: offices and laboratories. Plumbing, heating/cooling and irrigation requirements include a significant proportion of the normal consumption of office water. By constructing high-efficiency plumbing devices and reducing irrigation, the Agency has lowered these utilizations.

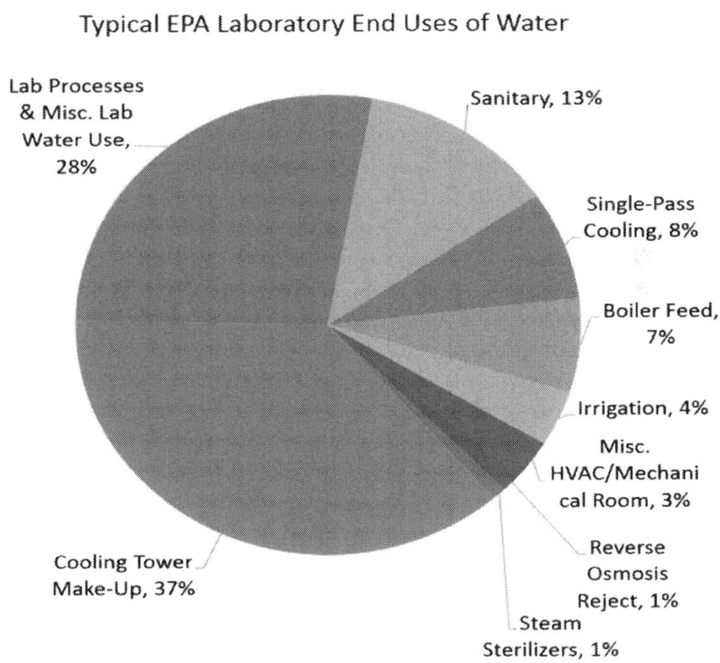

Figure 4.13. Typical EPA Laboratory end uses of water.

Source: https://www.epa.gov/greeningepa/water-conservation-epa.

Laboratories employ water to process the lab, purify water, and sterilize steam. Laboratories also have heat loads; therefore a considerable volume of water is employed as a aquilege tower for cooling. EPA has concentrated a lot of its efforts to save water on laboratory installations. In 2002, before any government rules requiring water conservation, EPA began to reduce its water footprint. The planning and reducing efforts in water management by EPA have yielded considerable results (Vyas & Nath, 2021).

4.11. STRATEGIES FOR EPA

The strategies for EPA water reduction include:

- Monitoring water meters and tracking use.
- Installing Water Sense labeled and other high-efficiency restroom fixtures.
- Eliminating single-pass cooling.
- Optimizing cooling tower efficiency.
- Minimizing or eliminating landscape irrigation.

4.12. CONCLUSION

In the modern environment when people can alter our planet's environment, piecemeal responses and reactions to unwanted interruptions in life support systems aren't enough. In this Anthropocene era, we are the only species capable of thinking and changing our behaviour to ensure our health as well as our economic and social well-being. Our key obstacle in this century is to provide a healthy and meaningful existence for all mankind. It will take all of us as a civilization to establish technological, economic, political and social strategies that will lead to achieving a desired and viable and sustainable future through research, development, technology, and science, and governance (van Koppen & Schreiner, 2014).

The research will always be necessary to find and assess the implications of alternate roads into this future and the compromises that are inevitable given our numerous ambitions or objectives that are not always compatible. However, it is equally important that the outcomes of research are properly translated into ways in which policy is relevant. For example, the results of research given in WRR articles need to be "translated" into a language that is relevant to policymakers and the public. It must be written in such a way that it encourages them, entertains them and instructs them.

Researchers from numerous fields need to work collaboratively in future. It's good to draw inspiration from the vision of the American Civil Engineering Society members view:

"Coordinates civil engineers, who work in partnership and ethically as:

1. Master planners, designers and building designers and operators of the economic and social engineering society, to create a sustainable world that enhances the global quality of life.
2. Environmental stewards and their resources.
3. Innovation and integration of public, private and academic ideas and technology.
4. Controllers of natural occurrences, accidents and other dangers and risk and uncertainty.
5. Pioneers in public environmental and infrastructure policy debate and decision-making."
6. The main problem facing all of us is how to reconcile our desire for a happy existence in this world with the restrictions imposed by the availability of water resources that are renewable but limited. You can do that. Let us be an optimism to illuminate the path ahead!

CHAPTER 5

MANAGING WATER IN URBAN SETTINGS

CONTENTS

5.1. INTRODUCTION

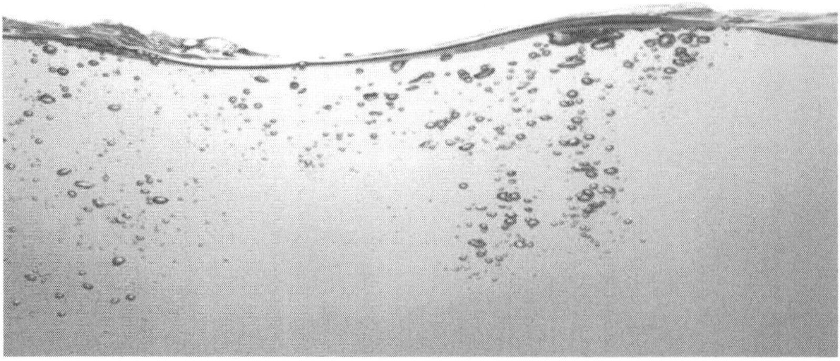

Figure 5.1. Water is an important resource that ensures the survival of different ecosystems.

Source: https://www.istockphoto.com/photos/water.

Water is an essential element for survival of different ecosystems. This is why water management is crucial. Water management involves different processes such as planning, development, distribution and optimum management of water. In urban settings, water management is crucial as water is supplied to individual homes and if not managed properly it could affect the functioning of urban settings. In most urban settings there are water management cycles to help ensure optimum management of water (Singh et al., 2021). Water is needed by all ecosystems to help both plants and animals survive. Plans made on water management are frequently changed or modified to meet the recent changing needs of water needs. With climate change being a major problem in current years, ministries in charge of water management in urban settings have to make proper management plans to ensure availability of water.

The need for water management has been brought about by the growing scarcity in of water. Researchers involved in this field say that there are two main phenomena involved in explaining the growing water scarcity. One of the phenomena is the growing need for fresh water while the other is decline in usable resources such as fresh water. According to the latest research, global water usage has increased more than double the rates of population increase. The data was a result of comparisons made between the current century and the previous century. This data also shows an increase in regions getting access to water. Some of these areas are arid regions. With a sustainable water supply, these regions can get access to freshwater for drinking. As much as this is good news, while the bad side is that it has resulted in a dramatic increase in water scarcity. This has resulted in some urban areas opting to get water from other countries.

From recent research done on water as a resource, it is expected that by 2025 some countries will be facing severe water shortages. Currently, about 31 countries are dealing with water shortages. According to data released by the UN, it is expected that about 4 billion will face the challenge of water shortage. In India, about 29 major urban settings are affected by water shortages. This is very worrying as it could lead to unequal distribution of water. These are among problems that can be tackled by implementing a proper management structure. If future generations are to get a For sustainable water supply to the future generation, proper water management systems need to be put in place (Singh et al., 2020).

Water allocation is among the water management plan. In urban areas, water allocation is extensively done so that people have access to water at their homes throughout the day. However, with the continuous changes in climate, it might be quite difficult to ensure continuous access to water. For this reason, water ministries have to establish long-term water management plans. Thus, makes the decision-making process quite difficult for water ministries handling urban areas. The continuous changes in climate have resulted in situations that governments have had a hard time handling and resulted in adopting other management strategies to ensure continuous water supply. These strategies also ensure very minimum setbacks. The need for water management planning is crucial in ensuring equal allocation of water to homes and industries. Ministries of water need to meet the water demand in urban settings. While meeting the water demand, they need to ensure a certain limited capacity of water should besupplied continuously.

However, water ministries are rarely caring able to meet water demand in urban settings. The challenge with water management is water sustainability in the present and future time. The current problem is that water as a resource is becoming more scarce. Thedifferent uses of water is further increasing its demand. The need for water management is greatly increasing. There is a great need for achieving a balance. A balance needs to be achieved between usage of resources and human needs to ensure sustainability.

5.2. SOURCES OF WATER IN URBAN AREAS

Figure 5.2. Springs and rivers are some of the water sources.

Source: http://www.amwua.org/blog/multiple-water-sources-ensure-sustainability-for-cities.

There are numerous sources of water used in urban areas are as follows:

Spring refers to the point at which ground water comes out from to the surface. At this point, pipes are placed and then water is availed for consumption. There are different kinds of springs. Some are seasonal springs while others supply water throughout. Depending on the rate of water flow, water management centers decide on how water will be supplied to homes. In some urban areas of developing countries, water allocation is not well-define and uniform. For such homes, homeowners may decide to establish a borehole or a well. Wells and boreholes are established in companies, residences and schools. In some countries, boreholes and wells are built and a piping system is established that supplies water to residential areas. Hand-dug wells and boreholes are usually found in schools, industries, religious areas, hospitals and commercial enterprises. Wells and borewells are cheaper source of water.

The natural sources of water include rivers and lakes. Water is directly abstracted and kept in reservoirs usually in form ofdams that arebuilt across the river. The process of abstraction involves directing the water from natural resource to an area where it can be treated and supplied to consumers. Rain water is also a source of water. In some countries, residential houses are built with gutters that trap the rain water and it stored in tanks (Afshar et al., 2015). They are a good source of water. Water obtained from wells and springs can be availed to consumers in treated or unrelated forms. Water obtained from the spring isdirected to water treatment centers to get clean potable water.The water sources are termed as improved water sources in the cases where they are built and utilized in a manner that it is free from contamination. Such water should be free from all contamination including fecal matter. However, surface water is not considered an improved water source. It is not improved as it is not safe for drinking. This means that once water acquired from the source, it has to be taken to a water treatment center and once safe it is available to consumers. Untreated water is usually referred to as raw water.

Different states make use of water obtained from other sources. Different projects have been initiated to increase the sources of water. For instance, in Colorado, the cities have made agreements with the SRP to provide water to the cities. This project makes use of eight dams, several canals and water reservoirs that obtain salty water to the valley. While in the reservoir, water is purified and mixed with ground water improving water sustainability and meeting water demands. Different countries have initiated water projects that use water from unusual areas, purify and treat it and afterwards supply it.

Some cities make use of reclaimed water. This involves wastewater being collected and being reused. Wastedwater is cleaned and treated before it is used. However, this water may not be used for drinking. Treated wastewater is used in irrigating parks, recharging aquifers or creating fishing ponds and lakes. Treated wastewater is referred to as a renewed source of water as water is recycled. Treated water is referred to as A+ quality. It is also used in gold courses, playgrounds, schools, campuses and HIA common areas. Water with grade A+ quality has been subjected to treatment and later on disinfected. The water is then tested for disease-causing bacteria and if clean it can be used. This water should not contain any form of nitrogen compounds as such components can contaminate groundwater.

5.3. IMPORTANCE OF WATER MANAGEMENT

Water management is an essential practice in urban areas. This process is crucial as water movement is properly monitored. This is all done to ensure the survival of various ecosystems and reduces the damage that occurs due to poor usage of water. Water management is also important for optimum usage of water. It reduces the cases of flooding and other events. One of the areas where water is used is agriculture. Water management ensures that water is efficiently used and enables farmers to make use of the little amount of available water (Singh et al., 2021).

Water management involves a variety of activities that are quite expensive. Some of these activities are budgeted drainage systems such as surface and sub-surface systems are analyzed. This helps water ministries predict the amount of water available for the public. With this information, they can properly allocate water to different sectors and also keeps a track of water available in reserves.

Theallocation of water to different areas depending on the purpose can be planned for efficient water management. And changes can be incorporated at some management activities such as the modification of withdrawal rates of ground water.

With regards to irrigation, it will be possible to track the amount of water present and how it will be properly utilized. The same goes for other areas where water is used. Water management also involves research and adoption of other methods that will be used in proficient managing water.

Data drawn from the research will help governments decide whether they should adopt other water projects to boost available water in reservoirs. It also pushes governments to look for other sources of water. For instance, more boreholes and dams could be established if required. Water management also helps countries know whether they are properly utilizing water as a resource. It also helps ensure the survival of different ecosystems more so aquatic life. Water management ensures that these animals species will survive and support future generations.

5.4. TYPES OF WATER MANAGEMENT METHODS USED IN URBAN SETTINGS

There are a variety of methods for water management which can be adapted by water ministries and other relevant organizations. In urban settings, there is a huge water demanddue to numerous industries, companies and residential areas requiring water for different purposes. Different urban settings use certain water management methods depending on the suitability of the method for their area. The different methods of water management focus on the main objective of the management process which is effective usage. Some of the methods used are as follows.

5.4.1. Rainwater Harvesting

As the name states, this method involves the collection of rain water and storing it to prevent it from running off. This method is quite basic as it was used traditionally where containers would be left outside that would collect some water.

Over years the method was enhanced to ensure maximum water collection and storage. Currently, most modern homes in urban settings have been fitted with gutters whose pipes direct water to tanks. Therefore, the water collection involves roof-like surfaces which is used for water collection.

The collected water is then redirected to aquifers, tanks, cisterns, reservoirs having percolation or deep pits like boreholes. Nets among other tools have been used in the collection of fog and dew (Sroufe, 2018).

It is important to know the difference between rainwater harvesting and stormwater harvesting. Rainwater harvesting involves the collection of runoff water from roofs while water from roads, creeks and drains among other land surfaces are collected in stormwater harvesting.

Harvested water has extensive uses including drinking if properly boiled, watering plants, taking care of livestock, domestic heating and irrigation among others. With proper care, harvested rainwater can be stored for long periods of time.

This would require water to be stored in dark tanks ensuring atmospheric light does not reach stored water. This reduces the chances of algae growing in water.

Some countries have made it mandatory for urban settings to harvest rainwater to enhance the groundwater recharge. To make it possible, the government through the ministry of water should make the law and further awareness about it among public.

Some countries are reaping from this management strategy by recording an increase in water levels and improvement in its quality. This has resulted in a law requiring new housing societies to practice rainwater harvesting.

5.4.2 Ground-Water Recharge

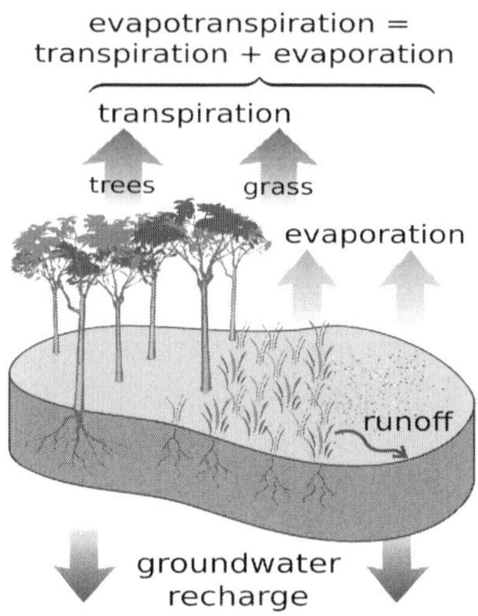

Figure 5.3. There are both natural and artificial methods of recharging groundwater.

Source: https://en.m.wikipedia.org/wiki/Groundwater_recharge.

Groundwater is among one of the sources of water in urban settings. Some of the sources falling under this category is springs. The process of enhancing natural groundwater is referred as groundwater recharge. This is achieved by utilizing man-made conveyances like trenches, injection wells, infiltration basins or dams. Among the common methods of groundwater recharge is Aquifer Storage and Recovery.

This method helps in to recovering groundwater and augmentation of groundwater resources to ensure water sustainability for future use (Alamanos et al., 2018). Groundwater can be recharged by natural methods such as snowmelt, surface water and rain. Climate change and nature, in general, could improve groundwater levels. The challenge with groundwater recharging islimitation brought about by human beings which slowdown the recharge to a great extend. Some of the human activities that impede natural groundwater recharge include logging, development and paving. All these activities cause the loss of topsoil leading to the decline in water infiltration. The lack of topsoil increases surface runoff while reducing groundwater recharge. In most cases, groundwater is used for irrigation purposes. If irrigation is not properly done it could cause a significant decrease in groundwater.

Water management mostly revolves around recharging groundwater as it contributes greatly to water sustainability. This can be priced by the fact that the water volume rate obtained from recharged groundwater is much more compared to the water volume rate gotten from an aquifer. Recharging of groundwater is essential as it reduces excess salts that tend to accumulate in groundwater systems, deep soil layers or root zones. The presence of tree roots reduces water runoff by increasing saturation into groundwater. Another method of groundwater recharge is temporary floods. It helps improve river bed permeability as clay soils move down the stream. It is also useful in facilitating aquifer recharge.

There are different methods of estimating groundwater. The variety of methods are a result of the difficulty in quantifying groundwater recharge rates. This is because numerous processes must be accounted while computing rates of groundwater recharge. These processes include transpiration, evapotranspiration, evaporation and infiltration. Physical methods are also employed to analyze the rates of groundwater recharge. These methods utilize soil physics principles in evaluating and estimating the recharge (Araujo et al., 2015). These methods can be categorized into two: direct and indirect physical methods. Direct physical methods revolve towards measuring the volume of water moving in soil zones lower than the root zone. Indirect physical methods are quite different as they are dependent on the estimation of physical parameters of soil. These parameters when used alongside principles of soil physics, are useful in estimating the likelihood of groundwater recharge. For instance, if rain is not experienced in a couple of months, water levels in rivers are low indicating little or no groundwater. This can be used in establishing the base flow of the area. The already established base flow is used in calculating groundwater recharge rates.

There are chemical methods of estimating groundwater recharge rates. This method uses inert water-soluble substances present in the soil. These substances are spread as soil moves into deeper layers. Examples of these substances include chloride or isotopic tracers. The final method of evaluation is numerical models. This method makes use of a variety of codes namely SHAW, UNSAT-H, MIKE SHE and Hydrologic Evaluation Landfill Performance. Some of these codes can be found online for eg. 1D-program HYDRU1D (Ashton & Turton, 2009). These codes make use of soil and climate data in estimating recharge rates. Richard equation is utilized to determine the flow of groundwater flow using modelling.

5.4.3 Artificial Groundwater Recharge

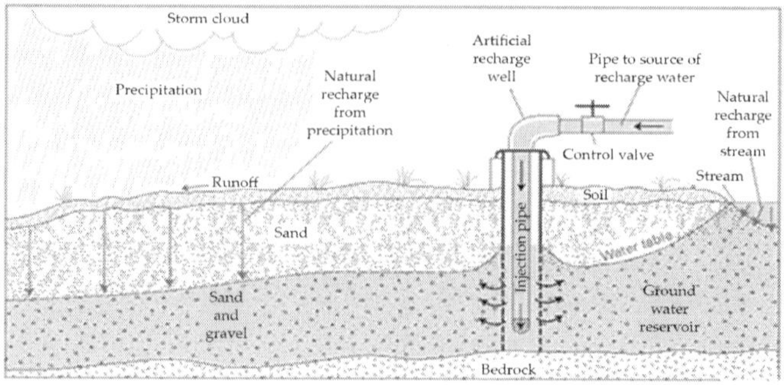

Figure 5.4. Man-made methods can be used to improve groundwater levels.

Source: https://www.usgs.gov/mission-areas/water-resources/science/artificial-groundwater-recharge?qt-science_center_objects=0#qt-science_center_objects.

Human activities impend natural groundwater recharge. This decreases the groundwater level in different areas. It will br devastating if the decline in water levels is beyond the rate at which aquifers can replenish themselves naturally. For this reason, artificial groundwater recharge is done. Artificial groundwater recharge is different from the natural methods, as man-made tools are used in improving groundwater levels. Human-controlled methods are used in increasing the amount reaching the aquifer. Canals, ponds and infiltration basins are used in redirecting water from land surfaces. This can also be achieved through the addition of sprinkler systems and irrigation furrows (Prinz & Singh, 2000). In some instances, groundwater levels are increased by using injection wells to direct water into the land sub-surface. In some urban settings, artificial groundwater recharge is very important as the excessive pumping of water for industries and domestic use has resulted in a significant decline in groundwater levels. For those urban areas, the government has to allocate funds to conduct groundwater recharge programs.

Among the factors affecting groundwater recharge is urbanization. Urbanization has both positive and negative effects on groundwater recharge. Studies have shown that groundwater recharge rates are usually higher in urban settings compared to rural settings. The high recharge rates is a result of the sewage networks and the extensive water supply in urban settings. Installation of the piping systems has resulted in soil being lost allowing water to easily penetrate through the soil. In urban settings, groundwater recharge is mainly through artificial means. Natural methods of recharge cannot work in urban settings as there are several road networks as well as infrastructure. Establishing such infrastructure makes infiltration of water to into soil difficult, implying that there are high amounts of surface runoff collected in dams or supply drains. The continuous increase in urbanization in different areas may increase groundwater recharge rates as compared to rural areas. Though this is good for water management, it alsohave negative impacts as it may cause flag floods. The increase in groundwater levels will require natural ecosystems to adapt to the change (Partha et al., 2019).

Natural groundwater recharge rates are reduced with construction of roads as the materials used in making them are less permeable. This increasessurface runoff. We can therefore say that urbanization plays a significant role in increasing groundwater recharge rates while minimizing water infiltration. This means that there will be frequent flash floods which will reduce as ecosystems adapt to the changes.

5.4.4. Drip Irrigation

Agriculture is one of the areas where water is extensively used. In farms, water is used to irrigate crops and take care of livestock. With regards to irrigation, it is important to manage water supply to allow consistent watering of crops. In areas where water is limited, drip irrigation is the best mode of irrigation. This form of irrigation is referred to micro-irrigation system. Other than managing water, this method of irrigation can save nutrients as water drips slowly to the roots of plants. There are different ways in which the pipes are placed. They can be placed below the soil surface or above the soil surface (Owusu et al., 2016). The main aim is to ensure that water gets to the root zone directly while reducing water loss through evaporation. This method of irrigation is said to be more efficient compared to other methods. The efficiency of drip irrigation is dependent on the design, installation, maintenance and mode of operation.

In a field, pipes are placed in certain intervals. The water flows through dripper lines. On the dripper lines, there are holes at different intervals and are referred to as drippers. Drippers are used for delivering nutrients and water to individual crops. Water is opened from the source and it contains fertilizers or nutrients. Water drops are emitted at each dripper. Once the taps are opened, there is equal distribution of water to each plant at a certain zone across the field. A network of lateral lines and sub-mains and man lined is used in delivering water to each plant. The drippers are of the same radius ensuring that equal amounts of water are delivered to each plant. This makes the irrigation controlled as nutrients and water are delivered at a controlled amount.

Figure 5.5. Drip irrigation is the most effective form of irrigation in water management.

Source: https://www.shutterstock.com/search/drip+irrigation.

There are several advantages of drip irrigation in water managementsuch as reduced loss of fertilizers and nutrients as water application is effective and minimum leaching. Compared to other methods of irrigation this method does not require levelled fields and can be used in fields with irregular shapes. This method of water management allows the utilization of non-potable recycled water. This method also allows the maintenance of moisture at the root zone, thereby requiring minimum amounts of water for agriculture. The frequency of irrigation is usually independent of the type of soil. It doesn't cause soil erosion and also reduce weed growth. This method of irrigation allows uniform distribution of water in controlled manner. This method of irrigation requires less labor as water supply can be regulated in crop fields by regulating drippers.

The disadvantages of this method of water management are; it isexpensive compared to other irrigation systems. The lifespan of the tubes is shortened by the sun as it heat melts it making it more prone to distraction (Olmstead, 2014). The sun can also tamper with the drippers causing some to expand. As the tubes are made of plastic, it is quite easy for plastic components to be found in foods and soil as a result of plastic degradation as plastic being exposed to the sun and different kinds of plastics making it brittle. The brittleness is a result of the flexibility when chemical compounds in plastics such as estrogenicis present. There is a high likelihood of pipes being clogged if water flowing pipes is not filtered. The pipes should be well taken care of and if not they can clog.

Water management is quite difficult for farmers who use a sub-surface type of drip irrigation. This is because they cannot see the amount of water applied which may result in excess water being applied and in other instances the limited amount of water is supplied to crops. This is common among those who are not familiar with the irrigation system. If not properly installed, this form of water management could lead to the wastage of water. While installing the irrigation system, several factors should be studied such as the topography of the land, agro-climatic conditions, soil, crop and water. One should also check the suitability of irrigation system to their fields. One should also consider the depth of installation depending on the type of soil. In the case of light soils, the installation should be modified in a manner that the soil can easily get wet with minimum water.

Generally, this method of water management is effective as it requires little amounts of water. Water can be saved in a manner and used for a long period of time. The method has proven effective and saved large amounts of water. In some instances, wastewater can e used to reduce the dependency of freshwater.

5.4.5. Greywater

Greywater is water that has already been used in showers, washing machines, bathroom sinks and tubs. Though it is used water, when collected it does not come into contact with faces. This is why water used in toilets or washing diapers is not collected. As this water is collected after different uses it may contain food, hair, traces of dirt, cleaning products and grease. Though it may appear dirty, greywater is safe for other uses. It is an advantage and provides supplementary water (Altenburger et al., 2015). Usually, water obtained from the household except toilets is collected through pipes and are directed towards surge tanks. While in the tank it is held briefly and can be directed to a treatment system or irrigation. Collected water can be driven to the surge tank by gravity if pipes are structured to flow with gravity or a pump can be used.

The surge tank is a large storage container that has been tailored in a certain manner allowing it to hold water in the early stages of the water surge. When greywater is released to treatment systems or for irrigation, the surge tank should be immediately emptied. This should be done every time water is released. Greywater should not stay in the surge tank for longer periods of time. If the tank is systematically placed; gravity can be used in emptying the tank. This is if there is enough gravity to drain the tanks.

In the case where greywater is to be taken to sewage treatment, it has to undergo a treatment process. The treatment process makes use of the source separation concepts. This ideology is used in different ecological satiation approaches. As mentioned earlier, greywater should not come in contact with toilet water or any source of feces. This is crucial as it prevents entry of loads of pathogens to greywater. This makes it easy to treat and reuse greywater. In the case where greywater gets in contact with wastewater from toilets, it is then referred as blackwater or sewage. Such kind of water cannot be treated in the same process as greywater would. Blackwater is taken to sewage treatment plants which uses a septic system.

Greywater generated from kitchen areas usually contains high organic matter content, fats, grease and oils. Such kind of water undergoes preliminary treatments to ensure that all contaminants are removed before greywater is taken to the surge tanks. In the case where preliminary treatments cannot be used then such water is taken to sewers and sewage systems. Greywater is much easier to treat and recycle as it contains significantly low amounts of contaminants (Ali et al., 2019). In the case where domestic greywater is obtained separately from the toilet water, it can be reused directly in homes or gardens. Reuse is depend on the owner's decision. It can first be treated or directly used. In the case where it has to be stored, the storage process should not for long durations of time. This is because organic components present in greywater may undergo a putrefy process. Though recycled greywater is said to be safe, it is not safe for drinking. Greywater can be reused as flushing or washing water. For this, it has to undergo numerous treatment processes for purification.

Greywater treatment processes are almost similar to those used in sewage treatment as they utilize the same principles. The major difference is the fact that greywater treatment is usually done on a small scale. Treatment systems may take the form of biological systems established in wetlands, normal backyards or living walks. The treatment system is small scale in nature and makes use of biodiverse landscapes or small ponds that act as a natural greywater purifier. It can also take the form of a membrane bioreactor or a compact system. Bioreactors are different from normal sludge processes. They can also be used in treating sewage. Greywater can also be purified through mechanical systems such as lava filter systems, treatment systems based on UV radiation or sand filtration. Contaminants such as food particles and nutrients found in greywater are used to grow constructed wetlands. However, contaminants such as soap and salt present a challenge to plant life and other microbial processes as they contain chemicals that may be toxic. These contaminants are degraded when taken to wetlands. The same is possible with the use of aquatic plants such as grasses, sedges and rushes.

Benefits of greywater is itseffectiveness in ensuring water management in urban settings. Greywater treatment plants may be found in many residential houses in urban settings. The use of greywater is reducing the demand for conventional water supplies and the need for sewage treatments. It is also advantageous for ecological watercourses and reusing of greywater lowers the amount of sewage waste entering these courses. Greywater has been useful in ensuring water management more during droughts. In urban settings, greywater has been useful in building for ecologically stable developments of such areas by re-using greywater in toilet systems and gardens.

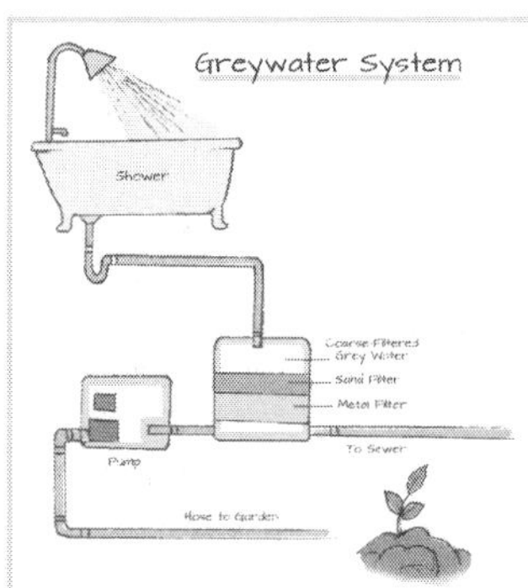

Figure 5.6. Re-using greywater has helped reduced dependence of extracted fresh water.

Source: https://www.google.com/search?q=image+of+greywater+sys5em&client=ms-android-transsion-infinix-rev1&prmd=ivn&sxsrf=ALeKk011BaBSBojluLrTW7zXkAwWAm2JYA:1623613163837&source=lnms&tbm=isch&sa=X&ved=2ahUKEwi434evrpXxAhU9QUEAHfFcDasQ_AUoAXoECAIQAQ.

Among the ecological merits of recycling greywater includes the reduction of the extraction of freshwater from aquifers and rivers. It also cut down the quality of wastewaterto be treated at treatment plants and septic tank infrastructures. It reduces the amount of energy used in wastewater treatment. It also minimizes the amount of chemical pollution caused by water treatments. It helps improve groundwater recharge and also improve nutrient reclamation (Badham et al., 2019). Natural purification of greywater is used in improving the quality of ground and surface water found in topsoil layers. The quality of water generated through natural greywater purification is better than that obtained from other water treatment processes. Greywater provides a great alternative to water management in urban settings in the Middle East. In such areas water resources are limited and this is worsened by the fact that the population in urban settings is growing quickly.

Other than providing alternatives to water sources, greywater has its economical benefits. As greywater has helped reduce the demand of freshwater, it has also led to the reduction of the usage of freshwater, leading to the decline in the cost of domestic consumption of water. This has been useful in reducing the global demand for water as a resource. This method of water management can help in the reduction of the amount of wastewater directed to on-site or sewer treatment systems.

5.4.6. Sewage Treatment

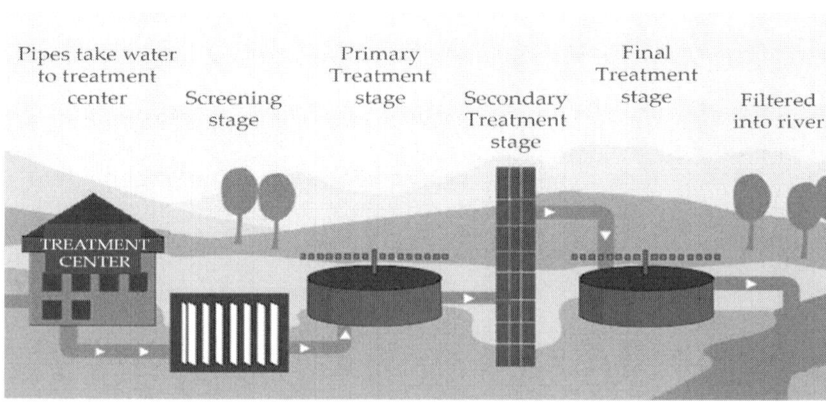

Figure 5.7. Sewage treatment involves the removal of toxic and solid wastes.

Source: https://eschooltoday.com/learn/sewage-treatment-process/.

Sewage is different from greywater in that its contaminants are more complex. As greywater can be immediately reused, that is not the same case for sewage as it is may have a very concentrated smell that isnot pleasant to all. The removal of contaminants from municipal wastewater is referred to as sewage treatment. In most cases, sewage is composed of household wastes such as those containing feces and industrial water wastes containing various chemicals. There is three main processes used for removing impurities from sewage water. The three processes are namely biological, physical and chemical processes (Olsson & Andersson, 2006). Treatment using three processes ensure the sewage water is safe for other usesand will not tamper with natural ecosystems. A slurry of semi-solid waste referred to as sewage sludge is a common by-product of sewage treatment. For sewage sludge to be properly disposed of or released to the environment, it has to undergo another treatment process. In most urban settings, there are sewage systems that will vary wastewater from industries and households to treatment plants. In most cases, industrial effluents are pre-treated in industries before they are taken to sewage treatment plants. This makes the treatment process much easier as the amount of pollutants is low.

Urban runoff tends to combine with sewage and increase a load of treatment plant. The runoff will be taken to sewage treatment plants. In urban settings without a proper sewer drainage system, the sewer may be exposed. Sewage treatment is a good form of water management as treated water can be used in various purposes and areas such as industries and gardens. The piping system may direct water towards a treatment plant. These pipes may direct water to flow due to influence by gravity are a properly mechanized pumping system. The main objective of the sewage treatment process is to ensure that there are little amounts of effluents reaching natural ecosystems. This means that the main goal is to minimize pollution. The primary form of treatment in treating sewage wastes involves the removal of solid materials. After the water is taken through a biological process that deals with the remaining soluble materialswhere flocs are generate to entrap fine solids. All the course and fine solids can be collected as sludge. The partially treated solid is free from solids and has reduced amounts of concentrated pollutants. The treatment process is made of three main stages they include the primary, secondary and tertiary treatment. To further treat water, there are intermediate stages and other polishing processes. There are also other methods used in water management processes.

5.5. TECHNIQUES OF WATER MANAGEMENT

There are several techniques of water management used in urban settings. They are as follows:

Measuring, Metering and Managing

This first technique performed after water being collected is measuring. In most urban settings there are meter units that measure the amount of water used in homes or organizations. From general calculation and analysis, one can look at the various saving options. The metering system is also ensures properly working water management equipment. This reduces cases of leakage and malfunctions in mechanical equipment thereby, minimizing wastage of water (Bagatin et al., 2014).

1. Optimization of Cooling Towers: Cooling towers are fitted in large industries, laboratories and facilities using large amounts of water. There are numerous operations done in cooling towers used in managing the ratio of water discharged to the amount of water lost through evaporation. Optimization of these processes helps in making results more accurate. The term cycle of concentration is refer to the ratio of evaporated to amounts of water discharged. To ensure high levels of efficiency, operations in the cooling towers should be at six or more than six concentration cycles. Metering is also involved in the process as it helps reduce cases of leakages and check for malfunctions in instruments.
2. Replacement of Restroom Fixtures: Different bodies have been involved in the managing restroom fixtures. Urban settings in Japan use re-used water in flushing water. In the U.S federal water-efficiency standards were established in the 1990s by the Department of Energy. All these bodies are working to ensure that toilet fixtures utilize a reasonable amount of water. For instance, some fixtures use about 3.5, gallons of water in a flush. This is very inefficient. By replacing these toilet fixtures in organizations, residential areas and companies, proper water management can be achieved. Some of the currently used fixtures include toilets having water flow rates of 1.3–1.6 gallons per flush, urinals flushing at the rate of 0.5 gallons per flush (lower for rates are recorded in other urinals), showerheads having flow rates of 2.0 gallons per minute and faucet aerators having a flow rate of 0.5 gallons per minute. These are not the only fixtures. Some utilize lower water flowing rates (Bouwer, 2000).
3. Elimination of Single-Pass Cooling: Before water is directed towards drains, it is taken through a cooling system to reduce its temperature. This is achieved through single-pass cooling. Most laboratories in urban settings make use of single-pass cooling. However, it was noted that it was not effective in water management causing the EPA to request for its removal. This has

resulted in laboratories opting to use recirculating or air-cooled chilled water systems. Many laboratories have eliminated the single-pass cooling system with recirculated chilled water loops. This is expected to ensure proper water management by reducing water use in laboratories by 80%. Some laboratorieswere able to save about 24.8 million gallons of water.

5.6. UTILIZATION OF WATER-SMART LANDSCAPING AND IRRIGATION

This technique calls for the planting of native and drought-tolerant plant species hence minimizing the amount of water directed towards irrigation. The native and drought-tolerant plants require little amounts of water. If an irrigation water audit is used, it reduces landscape water usage by 10–20%. Those installing irrigation audits should have the required licensing. There are irrigation audits that are weather-based while other use data on soil moisture to ensure that plants are watered if needed.

1. Controlled Steam Sterilizer Water: For steam condensed to be discharged from sterilizers to laboratory drains, cooling water is used to regulate temperatures. However, most laboratories use old sterilizer discharge that discharges a continuous flow of tempering water into laboratory drains. In this case, water is discharged even when it is not needed. The EPA has set laws requiring laboratories to replace the old sterilizers with those that have been fitted with tempering water control kits. The advanced sterilizers temper water when required. Laboratories that use the new model of sterilizers are expected to save about 1.5 million gallons of water.
2. Re-Using Laboratory Culture Water: In most urban settings, several laboratories use large amounts of water. For some, the large amounts of water go to aquatic culture research where large amounts of water are water from local bodies are pumped into specimen tanks in laboratories. To properly manage this water, it is taken to sewer treatment plants where it is treated and finally returned to the water body.
3. Reversed Osmosis System Operation: Purified water sold to consumers in supermarkets or taped in residential homes are result of reversed osmosis. Reverse osmosis is the process known to produce deionized purified water. Water management is achieved

by having proper regulation for purified water generation rates that meet the demand from laboratories. This means that systems used in the generation process are properly sized. By properly controlling reverse osmosis processes it is expected that companies will save about 1.5 million gallons of water (Nazemi & Wheater, 2015).

4. Recovery of Rainwater: Rain is among the sources of water. Most urban settings have established recovery systems to ensure proper management of rainwater. These systems capture water from roofs and store it in storage tanks. Recovered rainwater is used in cooling towers, irrigation gardens and landscapes or flushing toilets. In residential houses, recovered rainwater is used for domestic purposes. It can be used for drinking if it undergoes a purification process or is boiled. However, there is certain roofing systems used in the recovery of rainwater. This is because there are certain roofs made with toxic material which may contaminate the recovered rainwater. Recovery of rainwater has helped in saving about 300,000 gallons of water in a year.

5.7. INTEGRATED URBAN WATER MANAGEMENT

This water management project proposed in several countries seeking water sustainability in urban areas. The IUWM takes into consideration various factors needed to ensure its effectiveness. Some of the factors taken into consideration is all the various parts of the water system viewed as an integrated system, all uses of water, all consumers of water as well as the relevant stakeholders, local specific conditions and the need for water sustainability. Implementation of the IUWM is beneficial as it has resulted in an increase in water availability, reduced water treatment costs, improved biodiversity and opened the minds of water ministries to look for other water treatment solutions that are viable and not very expensive.

The process of achieving full integration of IUWM needs communication, collaboration and coordination across all relevant institutional boundaries. Multi-stakeholders should be involved. Successful results are gotten if there is cross-disciplinary interaction. While implementing, several processes are involved such as experimentation, learning and sharing of knowledge. As numerous experiments are done, regular reviews and adjustments are done on it. All stakeholders should share a vision for the project (Bouwer, 2002). The strategic plan for the project takes three approaches namely the scientific, action plan and strategy-systemic focus. The scientific approach oversees the observations, ideas, facts and solutions made with regards to the project. The action plan looks at the processes to be done.

While implementing the project, one needs to look at the enabling factors. For the IUWM some of the enabling factors include internal coordination in administration, stakeholder involvement and local politics. Under local politics is the laws shouldpromote the implementation of water management projects. Some of the information that has been used implementing these projects include the already available water resources, their sources, quality and quantity. Information on the rates at which water is used locally and expected water demands. Based on the water demand and the amount of water available, the project seeks to ensure that there is a balance between the two. This is why it may be searching for other sources of water. Some of the water management projects have led to the construction of dams and wells among others. The projects have also been actively involved in ensuring groundwater is recharged. They also look at areas where water is misused and they look for ways to minimize the wastage. They also emphasize on the need for water management strategies by highlighting their benefits.

CHAPTER 6

WATER POLLUTANTS AND THEIR SOURCES

CONTENTS

6.1. INTRODUCTION

Water is essential to the life of human beings and no one needs to be told of exactly how important it is to our existence. It acts as a cooling agent, as an ingredient of cooking, for farming among other things. The role that water plays is endless and there is no way one can say that they can survive without water. Nevertheless, individuals are not careful in terms of considering how they handle water (Butt & Bilal, 2011). Pollution is becoming an ecological threat and is seen to be affecting the existence of organisms starting from the life of the small organisms to human beings. Water pollution is experienced when there are toxic substances present in the water bodies such as lakes, rivers, oceans among others. The toxic substances are sometimes seen to dissolve in the water but some do not get to dissolve at all. The ones that do not dissolve either float on the surface of the water or deposit on the bed of the water body. All these deteriorate the quality of water.

Figure 6.1. picture showing pollution to the ocean.

Source: FairPlanet.

When this happens, it not only means that the life of the aquatic ecosystem is affected but also all of life, as we know it. Pollution not only affects surface water but also go deep into the groundwater. Pollutated groundwater reaches household which may cause the diseases.

There are many ways that may lead to water pollution and city sewage is one of the biggest causeshe. Making the cities to completely sewage free is quite difficult. As human beings, there is the process of excretion, which generates the wastes and it needs to go somewhere. Major problem is that most of individuals find it cheaper to release the sewage to the water bodies rather than find better ways to treat the same before releasing it to the water bodies (Brack et al., 2015). However, it should be understood that cheap is expensive.

As such, it may sound cheap to direct sewages directly to the water bodies but in the long run, the water gets affected, the aquatic life dies and human beings have no access to water of required quality. However, what exactly happens when the sewage is directed to water? Many years ago, individuals used to believe that the solution to pollution is dilution. What was believed was that when small quantities of the sewage was dumped to a large mass of water, then it would dilute by the process of dissolving. All this was believed to be happening through a natural process and what they called it self-purification.

The science behind this fact is true. As a matter of fact, during that era, this would be a very considerate process given as the waste that was discharged during that period was limited. With the industrial revolution, the amount of waste has increased and the water bodies do not have a chance of self purification with the high levels of pollutants.

In the current world, the amount of raw sewage is quite high and are very harmful. The discharge from the industries is very much noticeable nowadays. The smell and the color alone are unbearable. Just imagining the same going to the water that individuals use within the household and the effects on health is unexplainable. So one needs to consider treating their waste before deciding to release the same to the water bodies as it will have insane effects on the water that we consume.

Figure 6.2. Waste from industries directed to a water body.

Source: National geographic Society.

Nevertheless, what is raw sewage? Ideally, raw sewage is wastewater that has not gone through the treatment process. The sources of sewage are the kitchen, bathroom and even laundry. Consequently, the solid waste from the kitchen sinks, from the food that we eat, from dishwashing, the toilets among others also lead to the sewages.

All these lead to the deterioration of the water quality and further affects the aquatic ecosystem and as an extension to human life. Most scientists claim that in most foods that we eat, there is no need to throw away any part as each one of them has a role to play in our lives. Taking for instance potato peels which most people throw away.

The peels once washed carefully are very nutritious produce for human consumption. For the case of plastic substances that are used for preserving the food substances, the best use for the same would be to recycle them or if the ones that are currently in use cannot be recycled then they need to find ones that can be easily reused.

There are several things that can occur when raw sewage enters the water body. One is that the waste with lighter densitywill suspend on water bodies such as plastic and lighter substances that cannot sink. The other thing that can happen is that it can dissolve in water. This means that the water will have all the chemicals that have come from the sewage. The other thing that can happen is that the product can sink to the bottom of the water body and stay there. This is another aspect that has brought about a menace to the world.

The fact of the matter is that when these substances continue to accumulate at the bottom of the water body, the water level rises. This is where the fact that there are those water bodies are rising and going to areas that are used by human beings for their stay.

Apart from the fact that it destroys the water and makes it unfit for consumption, it goes further to chase people away from where they are staying. In simple words, it causes displacement of human beings. Ideally, the raw sewage contains impurities that are dissolved, accumulate and suspend. In the case of organic substances, it consist of food and vegetable waste. When the waste goes to the water bodies, microorganisms present in water start decomposing the materials and create anoxic conditions. This explains why sewages normally have a very unpleasant smell to individuals. The microorganisms are normally use the oxygen that is present in the watersfor their metabolism.

Figure 6.3. Food and vegetable waste in the water. Microorganisms come in to decompose the materials.

Source: BioEnergy Consult.

Certain microorganisms can be used for treatment purposes. But unwanted microorganisms create more troubles instead of being good to the environment. Raw sewage cause eutrophication of water bodies. In this process, more chemical is created. The levels of chemicals increase in sewage contaminated water which is not required for human life. (Cabrera & Custodio, 2014). Organic matter such as nitrates and phosphate among others are found in human waste, mostly serves as food for the algae and bacteria that are present in water. The increase in organic matter tends to increase the algae and bacteria that grows on water. This overgrowing of the same reduces the amount of oxygen present in the water. As a result, other organisms that live in the water get depleted of oxygen and in the process dies out because of lack of sufficient oxygen. The other organisms are strangled as they try to work hard towards survival.

Part of the problem is that the fish dies while the algae bloom. In addition to this, the pathogens that are present in the raw sewage can lead to the contamination of the water bodies making the water unconducive for human consumption and also for animal use. There are times when one can see the water corroding a piece of cloth, which is quite hard. Continous exposure of human to contaminated water may cause to many chronic diseases under compromised state of immunity. Ideally, raw sewage normally contains bacteria and viruses and other chemicals that are harmful to human, shellfish, fish, animals and other creatures that live within the water environment. Sewage is a composition of several pollutants which can be classified as explained below.

Figure 6.4. Sewages are the major cause of water pollutants since in most cases are directed to the rivers before treatment.

Source: The conversation.

6.2. ORGANIC POLLUTANTS

Organic pollutants normally include the pesticides and herbicides that are used either in farming or in pest control. One pesticide that is very effective in controlling mosquitos is Dichlorodiphenyltrichloroethane (DDT). In 1972, the pesticide was banned from the United States but the, its metabolite is found in the blood of most populations.

Other pollutants are polychlorinated biphenyls (PCBs), phthalates, dioxin that are by-products of manufacturing other substances. Phthalates are the plastic substances that are used in the manufacturing of many plastic products such as toys and other personal care products are used in hospitals. PCBs represents a larger group of pollutants that has different variations in toxicity and in most cases considered to be very persistent in the environment.

Some forms of PCBs are very similar to dioxins. There are compounds that are normally added to consumer items and can be detected in most populations, which are considered as fire redundant. All the above stated organic pollutants can kill individuals slowly (Carr et al., 2012). This is because these substances can be stored in the fats of human beings for years and may affect an individual once their metabolism proves to be weak. In most cases, it happens in old age when one has been affected by otherchronic disease. PCBs tend to affect the growth of human beings. A child that has been born from a mother exposed to smaller amounts of PCBs are normally seen to exhibit growth deficits. There are studies that claim to decrease the weight of fetal if a mother has been exposed to PCBs.

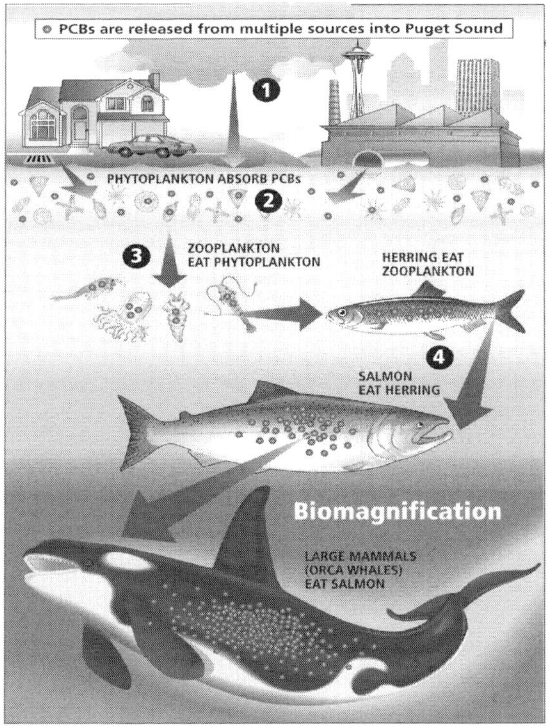

Figure 6.5. PCBs are found in fish. They are transferred from the small fish to the large fish.

Source: EnviroForensia.

PCBs are organic pollutants that tend to resist the metabolism process. They contain several chemically related and fire-retardant compounds. Most forms of chemicals degrade slowly in the environment and are known to affect the nervous, cardiovascular, endocrine, immune, reproductive and endocrine systems.

Overexposed to organic chemicals such as PCBs are carcinogenic and tends to bring about atherosclerosis and inflammation. In most cases, PCBs is responsible for skin and eye associated problems. Various techniques have been proposed by scholars to deal with PCBs in the fish oil refinery. High-temperature deodorization is one way of removing PCBs.

Absorption of the monopolar adsorbents such as activated carbon is another way of removing PCBs. The most effective method that has been put to use is the stripping of the chemical under optimized conditions where the environment is extremely hot and have high pressure. The problem with this method is that in the process, healthy fatty acids also are removed. Even though these methods are said to be effective in the removal of PCBs, they are not that effective in removing all the organic toxins that may act as pollutants in water (Muller, 2015).

The effects dioxin are also dire and considered as organic pollutants. It is produced as by-products during manufacturing of pesticide or herbicide, hospital waste incineration, pair bleaching among others.

It is normally described as a reproductive toxicant as it affects the reproductive system. In most cases, it alters the levels of hormones, which hampers the developmental process.

Exposure to dioxin normally occurs through food substances such as dairy products, meat or fish. There is no concrete solution available to reduce this kind of pollutants so far and the researchers are trying to come up with ways of curbing such problems.

In most cases, organic pollutants that are in water tend to accumulate in shellfish and are then transferred by the same. Organic pollutants tend to accumulate and increases the solubility with increase in the fats content of shellfish.

When there is a decrease in the number of pollutants in the environment either by cessation of the input or the transfer to shellfish can improve the quality of water but these pollutants can again enter the water. The depuration process will continue as long as there is decrease in the amounts of pollutants that are released to the water. In most cases, organic pollutants very common and causes major impacts on the lives of human beings.

Organic pollutants tend to contain molecules that have carbon in the third composition. Methyl tert-butyl ether (MTBE) mostly known organic pollutant (Ludwig et al., 2014).

Initially, MTBE was used as an air-cleaning gas additive. Currently, the chemical has been banned from use but it will take years to completely remove it from the environment and most importantly the water systems. Water that has been contaminated by this substance can cause leukemia, lymphoma, and tumors to the thyroid glands, kidneys and testicles.

6.3. PATHOGENS

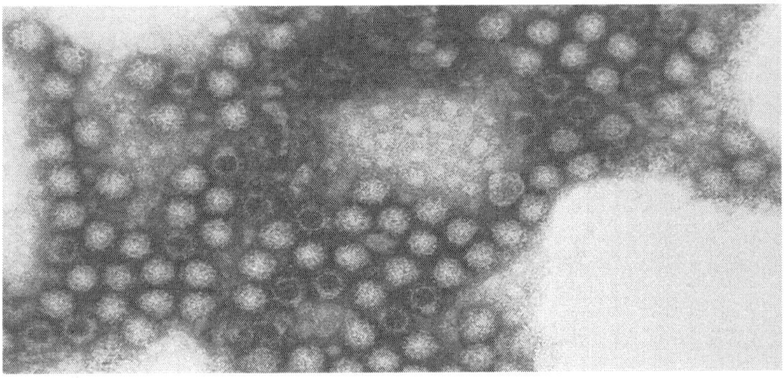

Figure 6.6. Pathogens are found in water and in large amounts are harmful to the water.

Source: Medical News Today.

Bacteria, protozoa and viruses all fall under the category of pathogens. Bacteria are commonly found in water and are used in the sewage treatment process. These pathogens are commonly found in human and animal waste and are in most cases the reason for diseases especially through the route of contaminated water. Ideally, pathogens are found everywhere including the developed countries as they can accidentally find their way to water bodies. Runoff from farms also brings harmful pathogens to the water bodies that are seen to affect human beings among all other organisms. The diseases that are normally spread through unsafe water include cholera, giardia, and typhoid. The United States is considered as a wealthy nation but more people are affected daily by these diseases thus indicating the severity of environment pollution (Lopez et al., 2009). Some individuals from the developed countries are using bottled water for their young ones to protect them from any contaminating diseases. Taking for instance in flint, the residents are facing crisis and the nation is trying to come up with contamination measures. The problem however goes beyond flint and is rather much bigger out there.

For the economy to thrive, all the aspects of the ecosystem play their role. That means the biotic and the abiotic components play their varied roles and when one of them is lacking, the entire ecosystem will fail. In this case, water is the part of the ecosystem that is failing and as a result, the entire environment is paying the price.

Microorhanisms are safe and needed in the ecosystem in order to complete the cycle. However, the problem occurs when their number rise above the safe levels and this lead tocontamination of the water. The most common pathogenic bacteria are coliform such as the *E. coli* bacteria. These bacteria are normally found in the environment and in most cases are considered safe organisms. They are indicator of the contaminated water and normally, water is analyzed for the presence of coliforms using the biochemical tests. Nevertheless, when these bacteria increase in number, they are quite unsafe for the ecosystem. They are considered dangerous to the environment. When water has been contaminated with either human or animal waste, the *E. coli* bacteria will be noticed in the water.

6.4. INORGANIC WATER POLLUTANTS

Inorganic pollutants are discharged from chemical and related industries, which include refineries, fertilizer industries and big or even small pharmaceutical companies. Heavy metals and some of the other trace elements, mineral acids, metals that have organic compounds as complexes, cyanides, metal compounds, metals, and inorganic salts. A high concentration of inorganic pollutant above the limits set by the regulatory bodies will essentially pollute the water. Some of the inorganic pollutants are biodegradable but most are non-biodegradable which means that they would persist in the surrounding environment (Leahy & Anderson, 2008).

Many of these inorganic pollutants can disrupt the health of individuals and also affects the flora(plants) and fauna(animals) that are found close to, on or in the water. In the recent past, there have been studies and research on the field of inorganic pollutants and they have mainly concentrated on their impact on the environment. Water is essential for life and living organisms cannot survive without water, which simply means that the availability of safe and pure drinking water is a basic human right. In the last few years, there has been a lot of waste discharged from industries, companies dealing with construction, and those dealing with sewerage disposal contain a lot of mineral acids, trace elements and some heavy metals that destroy and deteriorate the quality of water resources.

All these are mostly brought about by industrialization and rapid increase in population especially in the 20th and start of the 21st century. There are inorganic contaminants like arsenic, chromium, manganese and iron are often found and occur naturally in the environment. These metals and elements are often found on the surface and groundwater. Certain inorganic pollutants occur at high concentrations due to influence of human activities. These inorganic pollutants will affect the odor, color and taste of drinking water.

Figure 6.7. Arsenic is semi-metallic material found in groundwater. It appears like a metal but is not metallic but possess metallic properties.

Source: medical news today.

1. *Arsenic*. It is a naturally occurring component and part of the earth's crust and it is one of the most widespread contaminants in the natural waters. Arsenic is a natural and semi-metallic chemical and is found mostly in groundwater. Arsenic is a metalloid and it is not a metal but share most characteristics with metals. Research shows that arsenic is not consistent in all geographical areas but can be high in certain areas and not be present in others. This condition results from activities such as mining, application of pesticides etc. It can be found in a combined state with other elements as it can also combine with different compounds. Arsenic can be released in water by several human activities such as mining, found in animal feed, as a wood preservative and sometimes in chemicals used to control pests, especially in agricultural farms.

   Arsenic is odorless and tasteless. It is hard to identify arsenic water contamination by visual examinationc. It occurs naturally mostly on the earth and in the sea as a result of mining gold, lead and the combustion of coal. Some industries release their chemicals into the water and most of them use these compounds and release directly into water sources. Naturally, arsenic is thought to form certain type of rocks especially when groundwater drops significantly. With the rise in the water again it carries the arsenic. Surface arsenic will enter the groundwater systems by moving with the flow of groundwater from other sources such as rain and melting snow.

Inorganic arsenic is very harmful as compared to organic arsenic. Organic arsenic is found in seafood and the body will readily remove but inorganic arsenic cause diseases and harm the body (Larson & Lach, 2008).

2. *Silt*: This is common in areas that have bare lands and areas where water generally settles. It causes siltation which is called as siltrification. The silt or clay particles accumulate or concentrate on bottoms of water bodies and this especially happens in dams, the entrance of water into the oceans, lakes or seas. In most instances especially in rural areas silt is formed.

Figure 6.8. Siltation process where fine and loose soil settles in waterways and large water bodies.

Source: Wikipedia.

Soil degradation caused by intensive agricultural practices that are harmful to the environment. This then leads to the loosening of the top part of the soil. When the rainy season begins, the loose soil is carried away. In urban areas, there is a lot of construction and as it happens, the soil is loosened. The vegetation covering the soil is removed before constructing on land \ withand result in carrying away the fine soil particle by rainwater.

Dredging is performed in water bodies, especially along the coasts where industries and government have to build their ports. In order to construct, they have to dig to make the water dipper for big ships. In some countries such as the United Arab Emirates in the city of Dubai to be specific, they are carrying out a lot of dredging to construct artificial islands. While other reason of carrying out dredging is to remove unwanted materials from the shores (Cook & Spray, 2012).

Silt is produced bymachines such as vehicles operate in wet conditions especially wet roads. This makes the soil fine and easy to be carried by water.

3. *Acidity of water:* This is mostly caused by discharges from industries. Water can be acidified by various compounds. Researchers have found that along with the major shipping lines sulfur dioxide iss emitted from ships cause acidification of the ocean. As compared to areas where ships did not pass, there was no acidification and if present, it was not as prevalent as where the ships passed. In some instances, some shipping companies use an exhaust gas cleaning system in order to avoid air pollution and improve the air quality as per the standards set by the European Union. The water that is used in the cleaning systems contains sulfur dioxide and it is released into the water leading to acidification.
4. *Fertilizer:* This compound is used to enhance food production. In the recent past, most farmers have been using fertilizers made from industries instead of relying on organic fertilizer. These fertilizers are applied by spreading them on top of the soil. They contain a lot of nitrates and phosphates. The fertilizers dissolve in the rain water andleach into the soil and sometimes it reaches the groundwater. When there is a lot of rain and the soil becomes saturated with water, the water will carry away the nitrates and phosphate into rivers. Farming at steep locations cause movement of fertilizers very easily. Some nitrates and phosphates are supplied to plants in liquid form and hence have to be sprayed and in most cases, they will evaporate and they will form acid rain and this will be deposited in lakes and oceans.

Some industries use a lot of different chemicals andreleased most of them into rivers. These wastes are generated from manufacturing processes and other activities within the company. The wastes range from, garbage from the cafeteria which contains food and other chemicals used in the cafeteria for food production, chemicals for cleaning utensils, scrap metals, cleaning agents, i.e., for washing the machines and many others (Kirschke & Newig, 2021). Many industries have long pipes that carry their wastes from the industry to the point of release. Many are required to treat the waste and release them to the environment when it has been rendered harmless. Many industries by-pass regulation and release the wastes directly into water sources such as rivers lakes and oceans.

Figure 6.9. Vehicles produce exhaust that contains lead which sticks onto the tarmac and it is carried away by rainwater and into the storm drains.

Source: Cosmos magazine.

Heavy metals such as lead, is released by vehicles. These metals are found in vehicle fuels. After the fuel has been burnt and necessary energy released by the engine, wastes come out of vehicles come in form of smoke contains lead. In most towns, there is runoff as the rain water urban areas carries trash, heavy metals, gasoline, oils and other materials from car parking lots and roadways. Due to the impervious surfaces in towns and urbanization, the water is directed into dams or rivers.

Creosote is an example of carbonaceous chemical. They are mostly used as preservatives or antiseptics. It is mainly used in the preservation of wood. The creosote can readily dissolve with water. Wood that had creosote is sometimes released into the water.

Copper and iron eneter water bodies from acid rock drainage. Acid rock drainage is a process that mostly occurs naturally and is part of the weathering process in rocks. This is mostly a slow process but it is hurried by large-scale disturbances of the earth surface. In areas where liquid that comes out of the rock surfaces contains a lot of toxic substances (Krampe, 2017).

6.5. MACROSCOPIC POLLUTANTS OF WATER

These are large visible items that are found in waterways or small and large water bodies. The most common trash which has been dumped intentionally or unintentionally is washed away and find their way in water bodies through runoff and storm drains. The formation of the great pacific garbage patch which is now almost the size of France. Some other examples of macroscopic pollutants include small plastic pellets, pieces of metal and pieces of wood. There are shipwrecks and shipping containers. Although it looks manageable, individuals are assuming these pollutants.

Trash or garbage is the most common type of macroscopic pollution. Tash is essentially all materials that are used by human beings and later dumped. There individuals after having snacks dump the wrappings onto the ground. Due to large populations in cities, towns, and urban migration, many people live in cities and due to this, they dump their waste outside with no concern. The individuals throw the trash outside as soon as they finish using any item when it is of no use. People living along rivers, usually dump the plastic papers and kitchen waste both biodegradable and non-biodegradable into rivers. This is because they know rivers are moving water bodies and will not be stagnant hence there will not be any smell from the waste. After dumping the waste into rivers, it is carried into larger water bodies where they settle.

After individuals dump the materials on surfaces, the rain comes and the rain water carries the plastics. The water in most cities is carried by storm drains and trenches along the sides of the road. When the water is directed into the drain, it carries any material along its path. The water flows directly into lakes, dams, rivers, seas and oceans.

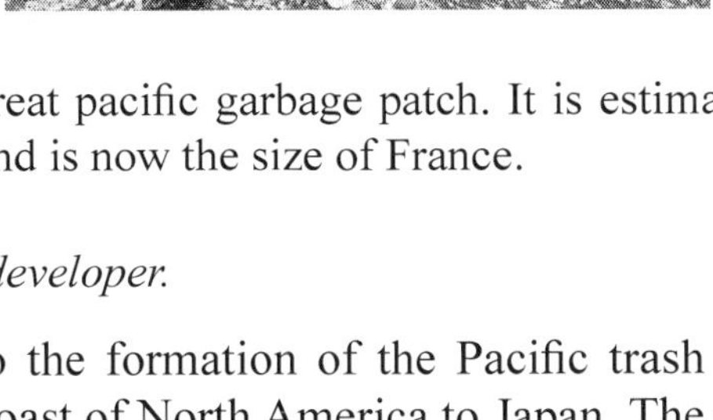

Figure 6.10. The great pacific garbage patch. It is estimated that the plastics have accumulated and is now the size of France.

Source: The urban developer.

This has led to the formation of the Pacific trash vortex. This spans between the west coast of North America to Japan. The patches are divided into two; the western garbage patch, which is located near japan, and the eastern garbage patch, which is located near the United States state of

California and in Hawaii. Origin this trash is from the countries close to the Pacific Ocean. This trash seems to be very stable as at the center of the gyre it is calm (Dobriyal et al., 2012). The motion of the gyre is circulating hence the trash and garbage being at the center. An example a bottle that has been dumped on the shores of the Pacific Ocean in California, it is first taken by the California current and then to the south towards Mexico. It then catches the north equatorial current which crosses the Pacific Ocean and moves towards japan, the bottle can travel north towards Kuroshiro current. At the latter stages, it travels eastwards by the north pacific current. The vortexes that are gently rolling of the eastern and western patches pull the bottle towards it. The patches have been growing over the years and currently, it is estimated to be the size of France.

According to oceanographers, the trash that can be seen on the surface of the ocean is not the only trash present. They believe that the trash is also on the floors of the ocean. The denser trash such as wood, metals etc. sinks into the ocean floor. Land accounts for up to 80% of the trash and the remaining 20% is from marine activities. There are many different types of trash but the most common is plastics. This is because plastics are more durable and the malleability makes the plastics more and more being used and also its high utility. Plastics also do not biodegrade but rather they break into small pieces.

Nurdles- This is a type of plastic debris that originate from plastic particles that are used in the manufacture of large-scale plastics. They are the largest contributors to marine debris. In a study carried out in Orange County beach, researchers found the nurdles to be the most common contaminant on the beach. The nurdles that are waterborne can be raw materials for the production of plastics or from large plastics that have been broken down. Some plastics escape during the production process and they move into waterways or large water bodies. Most commonly, they are found inas sediments and beach. The two most common types are polypropylene and polyethylene.

In a country like the United Kingdom, there are over 6000 companies that deal with plastic related products. In entire Europe, the number of companies rises to over 60,000. In the year 2018, the production of plastics estimated to be around 60 million tons. In the entire world, there was the production of over 350 million tons of plastic in 2018 which when weighed is more then the entire human population. After their production, the pellets are transported across the world and during the entire production process from nurdles to pellet to the final product they are not cleaned properly and can easily enter the waterways such as oceans and seas. In the UK, it is estimated that there are up to 53 billion pellets, finding their way into the ocean every year which is close to 35 tankers dumping pellets in the ocean every year. The global estimate is around 230,000 tons in the world every year. Nurdles can enter the sea during transportation or processing. They are small and light hence they can be easily being carried away by air and end up in waterways (El Bedawy, 2014).

All the countries on earth use plastics and this means they have to locally produce or manufacture their own plastics. This has led to the massive production of pellets. When companies are using the pellets, they have to be processed and end up as final products. The final products after some time break down and dumped either by being cut into small pieces or through photo-degradation where the energy from the sun breaks down large plastics into smaller plastics that end up in rivers, oceans, dams and seas.

Shipwrecks are one of the largest macro-pollutants in our oceans. Goods and other commodities have to be transported across the earth from one continent to the other. The largest means is the ships but some prefer aeroplanes if the commodity is urgently needed. If there is no urgency, it is transported by sea. Ships carry almost 80% of all goods that are transported across continents.

Many container shipslost in the sea every single year. They get lost due to many conditions such as bad weather and the ships that have not been properly maintained and also some ships are too old to be allowed to ferry goods. With demands for goods from other countries increasing and the fact that there is the warming of the climate, this means no goods and ships have to be transported. The warming of the atmosphere causes more bad weather regularly and also there is more unpredictability. All these factors when put together affect ships and more ships will be lost at sea. It is estimated that almost 1,000 containers are lost at sea every single year. These available data but certain countrieswithhold their data, which means there are more containers than 1,000 that get lost annually.

Since 1994 there have been recorded that 16,000 containers have been lost at sea. Shipwrecks at sea could be intentional or unintentional. The ship is either sink to the bottom of the water or it is be beached on land. A reporter known as Angela Croome in 1999 reported that there are an approximately three million shipwrecks across the world. There have been many known ships that were wrecked for example; South America which was an ocean liner and cruise ship was built in America in 1940. It was designed by naval architect William Francis Gibbs. She carried different names as she did her service. She wrecked while she was named "American star" on the 18th of January 1964 after 24 years of service (Kim et al., 2015). She wrecked near the shore and she has since sunk and the bow can only be visible during low tides.

Figure 6.11. A wreck of the SS American. The ship wrecked along the shore.

Source: sometimes interesting.

The Costa Concordia was a cruise ship that was operated by Costa Crociere. She has other sisters. She was built in Italy. On the 13th of January 2012 at around 21:45 hours the Costa Crociere hit a rock in the Tyrrhenian sea close to the shore of Isola Del Giglio. This led to a tear of almost 50 meters on the port side of the ship's hull (Jackson, 2006). This then led to the flooding of the engine room, which led to power being cut from another part of the ship and other ship services were disabled. As the water flooded in the ship she moved backwards and towards the shore where she eventually rolled on her side and settled. People were evacuated but 32 of them died.

Figure 6.12. The wreck of the RMS Titanic. It is at the floor of the ocean, it is degrading, and materials are spread in the ocean floor.

Source: Pinterest.

The most renowned shipwreck of all time is the RMS Titanic, which was a British passenger liner and was operated by white star line. The ship sank on the 15th of April 1912 in the North Atlantic Ocean. It is reported that it hit an iceberg during her maiden trip from Southampton to New York City. It was the largest floating vessel during its time. Harland and Wolff were the builders of the ship in a shipyard at Belfast. The architect was Thomas Andrews who was a chief naval architect of the shipyard at that time but unfortunately, Thomas died in the disaster. The captain who was Edward Smith also died in the disaster. On the 14th of April 1912 approximately 375 miles away from the south of Newfoundland which was four days after departure, the ship hit an iceberg at around 23:40 hours ship's time. This caused the plates of the ship's hull to buckle inwards along the right side and this opened up five of her sixteen watertight compartments. At around 2:20 hours early morning, she broke into two and started to submerge with more than 1000 people on board. Most of them were men as most of the women and children were evacuated onto the lifeboats first. It had the highest death toll arising from an accident (Hirji & Ibrekk, 2001).

The wreck of the titanic was discovered 73 years later in 1985 by a Franco-American expedition and was sponsored by the United States navy. The ship has been split into two and it is degrading inside the ocean at about 3,784 meters below sea level. Of all the survivors in the titanic Milvina Dean who was, only two months old at the time of sinking died at the age of 97 in 2009. The shipwreck causes a lot of pollution because as it degrades, the materials drift across the ocean.

1. *The HMHS Britannic*. This was a ship among those run by white star line. It was launched just before the start of the First World War. She was designed to be safest ship after the incident in the titanic. On the morning of November 21st 1916, she was shaken by an explosion that was caused by a naval mine. It was close to the Greek island of Kea. It sank 55 minutes later and thirty people were killed. The wreck was discovered in 1975 by Jacques-Yves Cousteau.
2. Many ships have sunk in the ocean and some are yet to be found. The wrecks are rotting at the bottom of the ocean floor. This leads to the pollution of the water surrounding the wreck. Some submarines have gotten lost in the sea and with materials on board that cause radiation, it becomes very bad as they cause a lot of pollution and they are very deadly.

Farming: Agriculture accounts for almost 70% of water withdrawals from rivers, lakes, etc. The farms release a large number of chemicals, which include agrochemicals, drug residues, sediments, saline drainage and organic matter. With a rapid increase in population, there is an increase in the demand for food across the world. This has led to the growth in crop production and this has been achieved mostly through the use of pesticides insecticides and herbicides. The areas that are being equipped to carry out irrigation has increased over the years whereas the total number of livestock has increased three-fold from 7.3 billion to almost 24.2 billion units. Livestock production accounts for almost 70% of all agricultural use and 30% of the earth's surface (Fang et al., 2013).

Figure 6.13. An organic waste lagoon. They carry organic waste from animals. When it rains heavily, the waste is sometimes carried away.

Source: environmental protection.

There are high levels of nutrients that are being used especially nitrogen and phosphorus and these two are synthetic in nature or produced as animal waste by-products. There are animal farms also referred to as factory farms where they house thousands of livestock where they produce billions of gallons of waste per year. For example, North Carolina alone generates almost 10 billion gallons of waste per year. The waste from animals is kept in pits or in open ponds called lagoons. This waste often leaks when there are large storms and some even rapture. To control this, some farms, spread the waste onto the field but in this instance also the waste leaches into the soil and it may sometimes reach the groundwater. Animals houses are washed using chemicals to prevent the spread of diseases and also to keep the animals healthy. After they have been washed, the water is directed into rivers.

Chicken waste contains high amounts of phosphorus and nitrogen. Most farmers spread the manure in their farms but the rate of use by plants is slow hence they are left in and on the soil. When it rains, the manure is taken away by the groundwater and surface water. All this water has its destination in rivers or lakes.

Farming where there is hills and mountains and big slopes causes water to move along them at a very high speed. When farmers dig the topsoil, they expose the topsoil. When it rains heavily, the rainwater carries the topsoil and during the wet season, the rains are consistent and it occur daily. The soil carried by water will be deposited in waterways. The continuous flow of water causes the formation of gully and with this means, water will carry away large chunks of soil into the river (Figuères et al., 2003).

In irrigation farms such as rice fields where water is used in large quantities, fertilizers are added, pesticides are sprayed and this is done while the rice basins are flooded because the rice needs large amounts of water. After a while, the water is released directly into rivers due to lack of regulation, e.g., in Ahero rice schemes in Kenya.

Chemical pollution can be prevented very easily. Ideally, human beings need to change their way of living and make their environment a priority. This is because they depend on the same environment for their existence. One way of doing this is by ensuring that one will buy only the chemicals that they needed and not the ones they want to buy. In addition to this, one should ensure that the chemicals they purchase are made in a manner that will not harm the water or the environment at large. One need to mix the products as recommended by the manufacturer and in addition use alternative fuels that cannot affect the environment. In the case of pollutants that are in liquid form, the individual needs to ensure that they treat the same before releasing it to the water body. If possible, the manufacturer can decide on recycling the water that has been used to ensure that the harmful substances are not released into the environment.

Discarding substances may seem like a common aspect to most individuals but they need to understand that their lives are on the line. As such, before discarding substances, one needs to look at the instruction to see whether the material is considered as recyclable material or it is needed to dispose of the material in a certain manner. Instructions are very important, especially when disposing of materials.

CHAPTER 7

IMPACTS OF WATER POLLUTANTS

CONTENTS

7.1. INTRODUCTION

It is believed that water is life; a fact which has been proven in so many ways. For instance, a large percentage of all living creatures require water for different purposes. One, they require water for hydration; in human beings, water is needed to prepare different meals, for cleaning purposes, for irrigation of plants and planted meals among others. Both wild creatures and domestic creatures rely on rain water to support the growth of the plants and vegetation they feed on. Every aspect of all living creatures does involve water.

For eternity, water is needed to be in its purest form before it is consumed by all living creatures. However, among human beings, the percentage level of the cleanness of water does differ depending on the usage purposes. For instance, the cleanness level of water for consumption is a bit higher compared to the cleanness of the water used for cleaning purposes or irrigation (Ghosh, 2017).

Over the years, the safety level of water used for different purposes has been declining. Methods of waste disposal are slowly being neglected with each passing day. For instance, in our current generation, individuals are dumping most of their waste materials into water resources. The principle of dumping the waste materials in the right dumping sites is no longer being adhered to.

Figure 7.1. An image showing multiple types of water pollutants.

Source: https://www.quora.com/What-is-water-pollution-What-are-the-effects-of-water-pollution-on-human-health.

The increased disposal of waste materials at just any other place has increased water pollution in different locations. Materials that pollute water resources are referred as pollutants.

The water pollutants disposed in different water resources have made access to clean water almost impossible. Contaminated water takes time before it is purified, and depending on the process of purification, not all pollutants can be completely removed. The most common water pollutants present in water sources include human waste and waste materials released from factories and industries. Human and animal waste is contains pathogens which end up polluting the water resources. Organic matter such as the remains of vegetables and foods are also water contaminant when disposed in water resources, as their metabolic product can pollute the water. Sewage water from our residence is also categorized as a water pollutant as it consist of several chemicals. Plastic materials do also contaminate water if they are disposed in water resources (Grizzetti et al., 2016).

Water pollution affects all living creatures, as all living creatures rely on water for survival as stated earlier. Contaminated water affects habitats of rivers, lakes and oceans; it affects the health of human beings,wild animals, plants and crops.

Human waste accounts for 30% in water pollution and organic pollution takes up to 15% of water pollutants, and the plastic waste materials take up to 5% in the major water pollutants.

7.2. IMPACT OF WATER POLLUTANTS

Every human being across the globe knows the importance of water, yet, we do not pay much attention to keep our water resources clean. Some of us have gone to the extend of polluting our water resources with the mentality that, it will not affect us, forgetting that it will affect our loved ones residing in different parts from us; and in one way or the other, goods which contaminate the water will come back to us, affecting us indirectly. The main reason why water resources are easily contaminated by the waste materials because we simply dump the waste into them thinking that water is a solvent that easily dissolves components released into the water, making them part of the water (Geissen et al., 2015).

The water pollutants which have been and are still being released into different water resources are have affected the globe in so many ways; from premature death of people, to premature death of animals and also plants as they also rely on water for their survival. Water pollution has led to certain health conditions which are new and difficult to cure.

Water Pollutants are believed to affect all water resources, however, there are six main water sources that are negatively affected by water pollutants.

7.2.1. Groundwater Contamination

Groundwater, as the name suggests, is water resources which exist deep in the underground levels of the earth's surface. Most groundwater is stored in water storing rocks which are referred to as aquifers. Groundwater is obtained from the ground in two forms; first as natural, where, it naturally finds its way above the earth's surface majorly through springs; and second in artificial form, where human beings have to dig deep into the earth's surface till they reach the underground water level, and they connect pipes and machines to draw the water upk on to the surface.

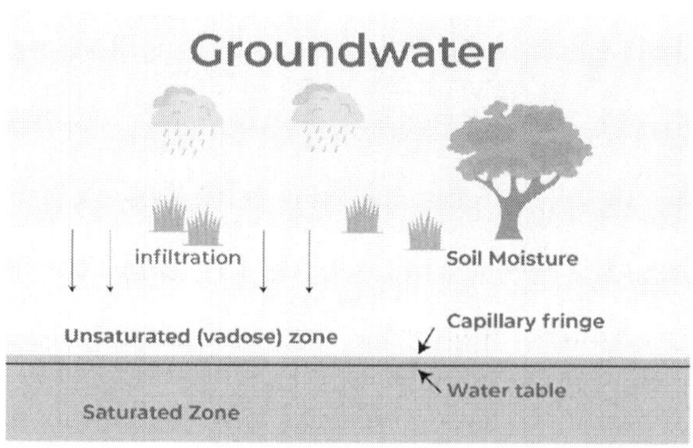

Figure 7.2. An image showing a groundwater level.

Source: https://www.trenchlesspedia.com/definition/2776/groundwater-table.

Groundwater get contaminated by pollutants in different ways. One, overloaded dump sites experience fast rates of decomposition; and with the decomposition process, harmful chemicals are released and continue sipping deep into the ground. More the decomposition rate, the higher the amount of chemicals beings released, increasing the rate at which the chemicals reach the underground water level. The chemicals dissolve in the underground water, and when extracted for use it is already contaminated. Excessive use of pesticides and fertilizers in farms increases the rate of underground water contamination. The excessesive pesticide and fertilizer solutions find their way into the underground water level, hence contaminating the water

(Fuhrmann, 2012). Water pollutants present in underground water hampers the quality of water making it unfit for consumption. The live formthat reside deep below the underground water levels also get affected upon contact with the contaminants. Furthermore, the contamination of groundwater makes it difficult to draw the water. As the water in aquifer hardens upon contamination with chemicals, making it costlier and difficult to draw the water from them. Contaminated groundwater can spread to other water resources which is a one of the major impacts of contaminated underground. This is because, the underground water level systems are interconnected with each other, and there are permanent water streams underground which supplies water to rivers, lakes and oceans. Hence, contamination of an aquifer underground water stream, leads to the contamination of all the other water sources majorly relying on that stream for water supply.

7.2.2. Surface Water Contamination

Surface water, as the name suggests, are all the water points that are found above the earth's surface such as lakes, rivers, oceans, ponds among others. Surface water can get easily contaminated since we have direct access to it.

Figure 7.3. An image showing a surface water source.

Source: https://www.americangeosciences.org/critical-issues/faq/what-is-surface-water-and-what-affects-its-availability.

Surface water can be categorized into two categories. The first category is freshwater which does not have any salinity components in it ; whereas the second category is saline water, which as the name suggests, has some salinity levels in it. Freshwater sources are the rivers and some lakes; whereas saline water sources include all oceans and a small percentage of lakes depending on the regions they have been formed in.

The major sources of surface water pollution are industrial waste materials, organic waste materials and artificial waste materials. Industrial waste materials make surface water impure for consumption. They also lead to the suffocation of surface water plants and animals, since the chemicals block or limit the circulation of fresh air within the water, making it hard for the aquatic organisms to breathe (Gao et al., 2014). Organic waste materials promote the growth of water weeds such as the likes of water hyacinth making it hard for aquatic animals and plants to breathe properly and also disabling some fun water activities such as surfing and boat riding. Artificial waste caused pollution of surface water are farm inputs such as the likes of fertilizers, pesticides and herbicides. If such farm inputs are used in excess, they end up into water resources, more so in farms that are close to the water banks, hence result in water contamination. The presence of such chemicals promotes the growth of water weeds, reduces air circulation in water sources as stated earlier.

7.2.3. Ocean Water Contamination

Ocean water, as the name suggests, is the water found in oceans and seas. It is believed that more than half of the earth is covered in water, and the large percentage of the water making up the earth's water is found in oceans and seas. Despite their large sizes, oceans are also impacted by water pollutants.

The major ocean pollution occurs at the shore or within the ocean. The shores have been turned into dumping sites, and also the waste accumulated in ocean are pushed back at shore by strong ocean water tides and they end up decomposing in the water releasing chemicals which affect the survival of water plants and animals in oceans. The washed-up waste materials also enetrthe aquatic animals, ending up negatively affecting them.

Oceans could also be contaminated by oil from the leaked vessels such as ships and boats. The oil leaks affect the survival of aquatic animals, as it limits the circulation of air within the water and also blocks the sunlight. Water birds that feed on sea animals and plants are also affected by the oil leaks by the clogging of their feathers upon contact with the oil during their dive in the water for food. The clogging up of their feathers makes it hard for them to fly that leads to the suffocation (Afroz et al., 2014).

7.2.3.1. Point source

A point source is a term used for a single source of water pollutants. Most point sources include the disposal points of waste materials from industries, oil refinery factories, sewage treatment factories and illegal dumping. Most industries release their waste materials in the form of liquids into the rivers contaminating the water making it impure for human consumption and also for the survival of aquatic living organisms. Most of the rivers, once famously known as the major fishing grounds have ended up being dead zone due to decreasing number of fish if any they are in small size.

Waste from oil refinery point sources affects the circulation of air within different water sources that in turn affects the survival of water animals and plants and also lead to the suffocation of birds of the air which feed on water plants and animals. Sewage treatment plants are not able to clean the sewage water completely, releasing it into natural water resources such as rivers and lakes. The consumption of improperly treated water affects the health of both aquatic animals and land animals including human beings.

Figure 7.4. An image showing a point source water pollutants.

Source: https://www.nationalgeographic.org/encyclopedia/point-source-and-nonpoint-sources-pollution/.

Point sources contamination might be viewed as a one region contamination although, it leads to the contamination of so many other regions, affecting not only a small percentage of animals and people, but a large percentage. For instance, industrial waste materials released into one point, leads to the contamination of almost the entire stream and also the other water sources which are connected to the contaminated water source.

7.2.3.2. Nonpoint Source

Nonpoint source water contamination, as the name suggests is a water pollutants source that is not specific. With non-point water pollutants, the waste materials come from different places. For instance, runoff water during the rainy season, carries so many waste materials disposed at sides of road, water trenches into the water sources, hence contaminating the water sources.

Soil debris from farms located in river banks does also contaminate the water sources. The soil debris contains different chemicals dissolved in them, once they fall into different water sources, the chemicals from farm inputs dissolve in the water polluting the water making it impure for consumption.

Figure 7.5. An image showing nonpoint water pollutant.

Source: https://en.wikipedia.org/wiki/Nonpoint_source_water_pollution_regulations_in_the_United_States.

Individuals travelling by sea or lakes do also cause nonpoint source water pollutants. For instance, the individuals disposing the plastic packaging materials of food into the water sources while travelling by water (Afshar et al., 2015).

7.2.4. Trans Boundary

Trans boundary, refers to water pollutants from one side of a nation that crosses over to the other side of another country. For instance, in oceans, an estuary located in one country might be piled up with plastic waste materials, which during the stormy seasons, are carried over to the ocean side which is located in a different country. The trans boundary water pollutants might also occur when a certain point of a river that crosses different nations is contaminated, and the water has to flow throughout the entire river, leading to contamination of other parts of the river located in different nations.

Trans boundary water pollutants affect not only one nation, but several nations leading to the occurrence of a pandemic due to the consumption of contaminated water.

Figure 7.6. An image showing a trans boundary water pollution.

Source: https://www.researchgate.net/post/Is_your_country_at_a_risk_from_transboundary_pollution.

Most trans boundary water pollutants are oil spills, disposal of waste materials from industries and factories, agricultural waste, farm waste and sewage water.

7.3. CATEGORIES OF WATER POLLUTANTS WITH NEGATIVE IMPACTS

7.3.1. Agricultural Waste

As it has been stated multiple times in the above sections that agricultural waste materials are among the top water pollutants affecting environment. Agricultural waste includes animal waste from farm animals, artificial waste from farm inputs such as fertilizers, pesticides and herbicides, among others. Animal waste once enter into water resources get decomposed by microbes which promotes the growth of water weeds and limits the air circulation in water sources. Farm inputs such as fertilizers and pesticides carried into water resources by runoff water during the rainy season dissolve into the water making it impure for consumption as it contains phosphates and nitrate in excess (Alamanos et al., 2018).

7.3.2. Sewage Waste

Sewage is generated by different activities in the house such as cleaning utensils, cleaning house, washing clothes, etc. make up the sewage water. Sewage water can also be generated by industries during manufacturing process and also from washrooms used in industries. Water running through the drainage system in big cities during the rainy seasons produce sewage. Since most rainwater enter the drainage systems due to impermeable surfaces and the water is mostly accumulate in different parts of the city has various contaminants.

Sewage waste is categorized as one of the major water pollutants across the globe. Most of the used up water is released back into surroundings without any treatment. Untreated sewage water once released into water resources affects the health of the human, the aquatic animals and plants.

7.3.3. Oil Spillage

Oil spillage is the other major water pollutant. Oil spillage is mostly common in oceans during the transport of goods. Some major rivers do also experience water pollution from oil spillage but the source of pollution is different. Rivers experience the oil pollution from vehicles along the road. Since the oils spilt on roads are carried into rivers by runoff water. Oceans are affected by oil spillages from the oil which spills from vessels which are used for transport purposes.

Oil spilt into rivers affects the quality of water. Oil limits the air circulation in water bodies, affecting the health of both aquatic animals and plants by hypoxia.

7.3.4. Radioactive Water Pollutants

Radioactive water pollutants release radiations above the expected amount and are quite toxic. Most radioactive water pollutants come from the dumping of uranium into river sources. Uranium reacts to release high levels of radiation making the water hazardous for all living beings. The radiations present water resources is harmful to the extend which turn a water resource into a toxic site.

Radioactive water pollutants are generally released from nuclear power plants. Most nuclear power plants use water for reaction purposes. In most cases, the excess water used for reaction purposes might be released back into the water source result in contamination of the water with radioactive substances. This is because the released water once used for reaction purposes will have some radioactive elements which with time will release radiations (Araujo et al., 2015).

Radioactive substances not only affect surface water but also groundwater. For instance, in situations where the radiations levels are so high either on surface, they might end up sipping into the ground contaminating the underground water making it highly contaminated and completely unfit for use. The excess radiations might find their way into the underground levels contaminating the underground water.

Water sources contaminated with radiations are hazardous to living being and cannot be used for any purpose; neither for consumption nor irrigation.

7.4. IMPACT OF WATER POLLUTANTS ON HUMAN HEALTH

Human health is impacted a greatly water pollutants. The most negative impact of water pollutants on human beings is that it leads to their death, more so if it has been contaminated with radioactive water pollutants.

Water resources upon contaminated, found to have a high load of unwanted chemical and dangerous microorganisms. Consumption of such contaminated water affect the health of human beings by causing major illnesses such as bacterial caused illnesses and in some serious cases, viral illnesses which are difficult to cure.

Contaminated water can cause illnesssuch as cholera, typhoid, giardia among other illnesses. The illnesses by contaminated water not only occur directly, but can also result indirectly.

Consumption of contaminated water may not always cause diseases directly. Mostly, systems supplying water to different homesteads, partially treated water, leaving behind some of the water pollutants such as the likes of pathogens, and some harmful chemicals in the water. Partially treated water appears clean but it may contain some harmful components which affect the health of the consumer (Ashton & Turton, 2009).

Most of the tap water in homesteads has heavy metals such as lead and copper dissolved in it. The consumption of such heavy metals leads to serious illnesses such as cancer to serious brain dysfunction issues, affecting the health of a human being greatly.

Human are social beings and engage in fun activities that allow them to mingle with their fellow human beings. Such fun activities might include water games and other water entertainment activities such as the likes of swimming. Engaging in Water activities contaminated water in one way or the other in most cases does affect human health negatively. For instance, swimming in contaminated water leads to the manifestation of pathogenic bacteria on one's skin, affecting the general health of their skin by causing skin rashes. External exposure to contaminated water can also lead to respiratory infections.

7.5. IMPACT OF WATER POLLUTANTS ON THE ENVIRONMENT

The effects of water pollutants on the environment have been briefly discussed in an earlier sections. Water pollutants have a negative effect on almost all components of the environment, either directly or indirectly.

It is believed that, for the entire environment to thrive and exist as one component, all the components which make up the environment should functioning properly as they are interdependent either directly or indirectly. The failure of one component of the environment may affects the entire system. For instance, if polluted water fails to support the growth of plants, other organisms which rely on plants as a source of food will also be affected and ultimately reduce in number, hence disturb the environment. Furthermore, the unavailability of plants affects the air circulation as plant consume the carbon dioxide in the atmosphere, balancing its percentages and release oxygen gas which is used for breathing purposes by many living organisms.

The water ecosystem is affected by water pollutants, such as oil spillages limit the air circulation levels and oxygen in the water, affecting the survival of aquatic animals. The growth of water weeds due to the presence of harmful chemicals within the water also affects the air circulation system within water bodies, slowly turning water environments into dead zones which cannot support any form of life. Furthermore, the water weeds do also release other harmful chemicals which affect all aquatic animals such as whales and water turtles, leading to their extinction in certain water bodies more so the oceans which are suffering from excessive dumping (Altenburger et al., 2015).

Some of the chemicals released into water bodies, like heavy metals such as lead and mercury, lead to excessive level affects aquatic plants. For instance, the heavy metals enter into the aquatic animals, affect the reproduction system of aquatic animals, making it hard for the animals to reproduce, hence the extinction of a certain aquatic species within a specific water body.

Plastic waste released into water resources affect the food chain in water bodies espectially in oceans. For instance, as stated earlier, aquatic animals might consume plastic materials as food that affect their body system and destroying their habitats. The decomposition of plastic in form of microplastic is also threatening the life of aquatic organisms.

The release of organic waste materials into different water bodies is affecting them by increasing the concentration of carbon dioxide within the water sources. Increased levels of carbon dioxide, more so in oceans further acidify the oceans making them more saline. Increased salinity levels affect the coral reefs where aquatic animals such as shellfish reside. Increased salinity also affect the survival of other aquatic animals, as most of them are not adaptive to extremely high saline environments making it impossible for them to survive and result in their early deaths.

Water pollutants not only affect the water environment or land environment but also affect the environment in which human beings reside. For instance, the presence of pollutants in water consumed at home, make the home environment too risky for the young ones the water is no longer safe, not only for consumption but also for other activity around with it.

7.6. IMPACTS OF WATER POLLUTANTS ON THE FOOD CHAIN

A food chain is referred as a hierarchy of organisms each dependent on the next as a source of foodand thus each organism rely on other for survival. Water pollutants affects the smooth flow of the food chain within an ecosystem.

In oceans, aquatic animals which rely on aquatic plants as a source of food end up lacking enough food due to the unavailability of plant life which has been disrupted by the chemicals released into the water bodies. The aquatic animals which feed on each either do not get enough food or consume other organisms which already have pollutants in them.

Reduction in the number of aquatic animals affects the human being as the large population is dependant on fish and other seafood for their survival. Thus water pollution affecting the water plant or animal directly or indirectly causing harm to human being.. The food chain in the land environment is also affected by water pollutants in different ways. For instance, the water used to irrigate different crop if contaminated may affect the productivity or quality of the crop. Almost all land animals rely on plants as their main source of food. Animals that feed on plants end up being limited in number due to the low levels of food within their ecosystem. The carnivores which feed on the plant-feeding animals, will also be affected due to their limited number. Human beings in turn end up lacking enough amounts of food from both plants and animals bringing about the issue of malnutrition on larger part of populatio (Ali et al., 2019). Water is truly life, and without water, it can be said that there is no life; a statement which can better be proved using the food chain disruption. Contaminated water affects the easy access to food to all living creatures.

7.7. IMPACT OF WATER POLLUTANTS ON THE ECONOMY

Water pollutants affect different aspects of human life. With water pollution, the economy is greatly affected in different ways.

One, water pollutants leads to scarcity of foods for all living creatures, both human beings and domestic animals and also wild animals. The scarcity of food reflects different negative aspects. The availability of labor work in different workplaces will be limited and individuals will not be strong enough to carry out their day to day activities, lowering the income made per day within a certain work environment. Scarcity of foods will increase the cost of basic food and its access to poor will be greatly limited.

Wild animals will also have survive on limited food. The scarcity of food supply to wild animals will cause death of a large percentage of their population. Tourism will be affected ascertain nations rely on foreign exchange collected from tourists who come to enjoy the beautiful wild animals; and with no wild animals, it means that the governments will not be able to collect foreign exchange hence the destruction of the economy. Water pollutants affect the economy as large amount of money has to be invested for cleaning up the contaminated water before it can be used for consumption by both human beings and animals (Badham et al., 2019). The collection of all contaminated water into one source is also costly as it requires the use of a quality piping system and quality storage units to ensure that the contaminated water is stored most safely and to ensure that it does end up leading to more contamination in other areas. The distribution of the treated water in other different parts of a certain nation is also costly. Furthermore, the already treated water has to be analyzed and confirmed that it is pure or not, it had to undergo another treatment phase, a process that costs money. As discussed in the above section, water pollutants are slowly turning the world into a world that does not seem to have a bright future aimed at water safety. We all do understand that water is life, and we need to truly value it. It is high time that the water pollutants be put away with. Strict laws and regulations should be implemented of industrial dischargedifferent water resources.

The water vessels travelling on water bodies have to be regularly serviced to ensure that they do not have oil leakage which end up contaminating the water sources. All sewage treatment plants ought to improve their water treatment technologies and ensure that they treat the sewage water in the most thorough way possible. All vehicles also ought to be regularly inspected and service to ensure that they do not leak oil which is later washed into water resources. Dumpsites should be regularly discarded to avoid the sipping of decomposing chemicals deep into the underground water levels ending up polluting the water resources (Bagatin et al., 2014). Farmers are advised to use just the right amount of farm inputs to avoid the sipping down of excessive chemicals deep into the underground water levels; and also to avoid the washing away of the chemicals into nearby water sources. Also the use of organic materials should be promoted.

Laws should be implemented tocontrol the irresponsible dumping of waste materials in just any part of their surrounding which will minimize the washing away of waste materials into water sources.

The future of our world and our survival relies on us. It is high time the water safety rules be taken seriously, and we all take part in ensuring that the water sources are not contaminated in any way, more so directly through our actions.

CHAPTER 8

MANAGEMENT OF CONTAMINATED WATER SITES

CONTENTS

8.1. INTRODUCTION

Menora Shirota's famous phrase is: 'Prevent illness instead of treating illness: a healthy intestine contributes to long lives and provides as many people as practicable with health advantages at a reasonable cost. This idea, developed about half a hundred years ago, is now more relevant than ever. There are numerous water treatment methods to combat the problem contaminated water supply. The technique used in this area, of course, is dependent on the features of raw water (i.e., contamination and extent), infrastructure (i.e., electricity, manpower, chemical supply), cost-effectiveness and acceptance (Bouwer, 2000).

Sedimentation or settling, boiling/distillation, chemical treatment (precipitation/coagulation/adsorption), disinfection and filtrating are some of the usual ways of water purification. The contamination mitigation processes and procedures are as such coagulation or precipitation is a method by which one or more chemicals are removed from a solution by adding reagents to produce insoluble flocs. The approach is regulated by 'solubility' the product of ion concentrations is precipitated (in simple form). It is one of the simple water purification procedures.

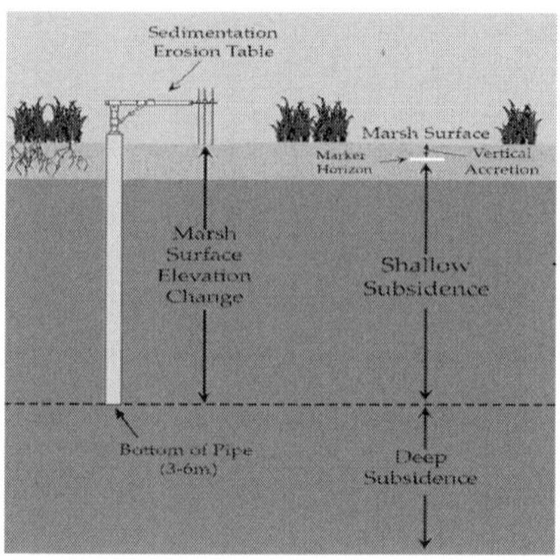

Figure 8.1. The Sedimentation Erosion Table (SET).

Source: https://www.researchgate.net/publication/228861064_Sediment_elevation_dynamics_in_tidal_marshes_functional_assessment_of_accretionary_biofilters/figures?lo=1&utm_source=google&utm_medium=organic.

The chemicals are added to water that generates particles which later settle down and the pollutants can be separated from the water. The purified water is used again while the settled section is washed away. The process is used for water softening and the removal of contaminants such as phosphorus, fluoride, arsenic, ferrocyanide, heavy metals, etc.

8.2. SOFTENING OF WATER

The carbonate, bicarbonate, chloride, Ca/Mg and sulfate presence increase the water hardness. The right chemical compounds can be used be to precipitate these componant to bring down the hardness of water. $Ca(OH)_2$ complements to bicarbonate and sulphate and precipitate them in water. Na-aluminate is added to hydroxide precipitation, including sulfate and chloride. Na-aluminate effectively produces sodium hydroxide in the water, and hydroxide forms through sulfate/chloride. Development of aids for floc formation, conditioning of slurry blanket and silica, reduction of aluminum hydroxide and easy boiling also makes water softening possible. Sulfates, sulphides, hydroxides and carbonates usually precipitate heavy metals (e.g., Ba, Cd, Pb, Hg, Ni, Cu). Co-precipitation of metals with iron and aluminum salts (e.g., As, Cd, Hg, Cr) is also feasible for some metals during flocculation. Chromium is co-precipitation with hydroxide or sulfate. Arsenic can be removed efficiently as coagulants, for example, alum $[Al_2(SO_4)3–18H_2O]$ and ferric sulfate $[Fe_2(SO_4)37H_2O]$. Arsenic(V) may be eliminated more efficiently than arsenic in other forms (Bouwer, 2002).

Figure 8.2. Morgan Salo, an engineer with Bolton & Menk, points to a tank where lime will be stored to soften the city's water at the new treatment plant.

Source: https://www.mprnews.org/story/2019/03/12/in-morris-other-cities-across-the-state-solving-salty-probem-in-municipal-water.

The flocks are attached with the microparticle and negatively charged arsenic ions by electrostatic fixation during the procedure. The possible steps of coagulation and co-precipitation are as follows: Alum dissolution, aluminum precipitation (acidic), co-precipitation (non-stoichiometric, non-defined product). Comparable reactions occur ferrous chloride and sulfate with the creation of Fe–As a precipitate as a final product, which is eliminated by the sedimentation and filtration procedure. The efficient elimination is pH-dependent. Phosphorus removal by coagulants, i.e., by mixing coagulants into wastewater, phosphates can be removed. Ca, Al and Fe are the most often used MMIs.

8.3. FLUORIDE REMOVAL

The most successful option for removing fluoride from effluent streams (as for Ca, Mg, Al) is the precipitation of fluoride in chemically stable forms which comes from research done by two scientists. CaF_2 is least soluble in the water among all metal fluorides. As a result of this, fluoride removal from the effluents most extensively employs treatment method that produce CaF_2. For this reaction, $CaCl_2$ limes can be employed, because of its greater solubility and lower additive-effluent ratio. The reaction by the aluminium is used as treatment agent for hydrofluoric acid and ammonium fluoride.: The separation of inorganic flocculants is different reflected by research, although they have very high amounts of utilization. This leaves vast quantities of sludge that are highly pH-affected, while polymer flocculants develop major cohesive aggregates (flocs) and least affected by pH (Butt & Bilal, 2011). For this reason, both natural and synthetic polymers are useful. In general, the plastic polymers are very efficient flocculent in tiny doses and possess great tailoring, but have low shear stability; although they are biodegradable and effectively stable in the form of natural polymers (i.e., starch, guar gum, alginate, Glycogen, dextrans). The teeth dyes are not biodegradable and difficult to capture thus $CaCO_3$ precipitation may be one of the best ways to remove them from the water.

The advantages of the precipitation technique:

- Easy to use.
- Suitable for Cd, Ba, Cd, Cr, Pb, Hg, Se, Ag, and so on for removing.
- The removal of natural organic matter (NOM) or organic carbon can also be implemented (DOC).

- Limitations:
  - Requires the ongoing provision of enormous chemicals.
  - By-products management.
  - Disposal of sludge coagulation is a concern.

8.4. DISTILLATION

It is the most popular approach of separation. The dissolved components in water are separated by using heat in this separation procedure. This technique is based on difference in the boiling point of different component present in water. The properties of the boiling point rely on the concentrations of present component. Therefore, the vapor pressure properties of liquid mixture rely on the distillation method. The fundamental principle of this method isthe heat energy increases the pressure on the vapor. The fluid blend boils and distillation take place because of the difference in the volatility of the components of the mixture when the vapor pressure exceeds the surrounding pressure. This procedure leads to the separation of water from inorganic elements such as lead, calcium, magnesium, etc. However, organic products with boiling points under 100°C cannot be successfully eliminated and get concentrated in the water of the product (Brack et al., 2015). Distilled water purification technology was originally developed for industrial purpose. It later used in household. Since, this technique does not remove organic contaminants very well, it is necessary to add a carbon filter system to make the water safe to drink. The carbon filters are changed regularly as they can quickly become a breeding ground for bacteria. While distilled water is safe, it is not healthy because it does not contain nutrients that are required for living being. This kind of technology is also quite sluggish to clean water. Furthermore, the cost and effect of a carbon filter is a manageable water purifying device.

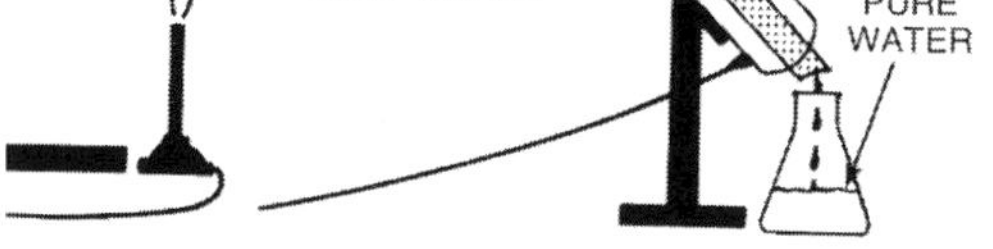

Figure 8.3. A schematic diagram of water distillation unit.

Source: https://earthandhuman.org/how-does-water-distillation-work/.

Benefits:

- Removal of a large spectrum of pollutants (toxic, heavy, bacterial, viral, pest).
- No physical barriers are reliable (filters).
- No further disinfection process is required.

Limitations:

- It consumes a huge quantity of energy in terms of refrigeration and heating.
- A condensate may include certain pollutants.
- Careful upkeep to ensure cleanliness.

It can be physicorption and chemisorption (originated by the forces of van der Waals). The adsorbent materials are added to the water supply directly or through a mixing pool. Adsorbents interact by chemical and physical procedures to pollutant and remove color, taste and odor components from the water. In principle, adsorbents is a microporous materials. However, the most favored ones are those with properly regulated and highly microporous. The porous materials, for example, activated carbon, silica gels, aluminas, zeolites, etc., have numerous cavities or pores as small as a fraction of a nanometer (Ali and Gupta, 2007; Qu, 2008). The isotherms represent a quantitative link between the adsorbing and adsorbing substances. Freundlich, Langmuir and Linear are the three most popular isotherms. Freundlich is the most widely used in water contaminants and expressed as follows: x m 1⁄4 KC1=n e: where x is the adsorbed solvent mass, M means the adsorbent mass, ce is the solvent balance concentration, and k and n are experimental constants.

8.5. CARBON ACTIVATION

Activated carbon is the most often used adsorbent, which is relatively similar to normal charcoal. Actually, due to its highly porous quality, the active carbon is considerably more effective. The high porosity character is achieved with or without oxygen in the presence of inorganic salts while processing carbon by steam and high temperature (130^{0}C) (physical method). Carbon can include petroleum coke, bituminous coal, lignite, timber products and coconut and groundnut shells.

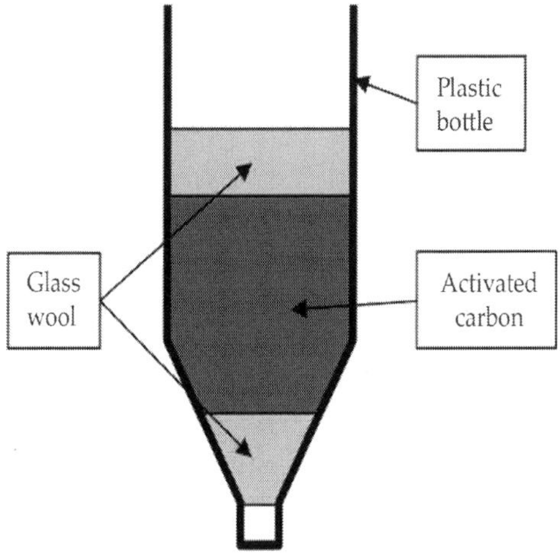

Figure 8.4. Design concept for the activated carbon water filter.

Source: https://www.researchgate.net/publication/234060484_Performance_of_activated_carbon_in_water_filters/figures?lo=1&utm_source=google&utm_medium=organic.

Carbon components are oxidized by CO_2 and steam at high temperatures (Cabrera & Custodio, 2014). The gasses are expelled and the carbon structure produces have micro fractures and pores. It substantially enhances the carbon surface area and makes it a valuable material for pollutant removal presented by research by scientists. In some instances, chemical activators such as phosphoric acid, zinc chloride and carbonated mixture can be handled at high temperatures and then washed away by washing water (chemical method).

Active carbon exploits the mechanism of physical adsorption, which enables them to attract contaminants onto its surface via Van der waals from the solution. The adsorption efficiency is dependent on the type of the carbon and pore particles, surface, density and strength, and contaminant type (concentration, hydrophobicity, polarity and solubility of the contaminant and contaminant attraction to the carbon surface). There are two types of activated carbon, granular carbon activated (GAC) and powdered activated carbon (PAC) which are commonly used. Both physically differ by particle size and diameter, as their names are implied. GAC can be mostly reused, because PAC particles are too tiny to reactivate.

Benefits:

- Active charcoal is excellent to the capture of carbon-based pollutants, chlorine (including cancer-based trihalomethans) as well as colors and fragrances.
- Extremely economical.
- Long life(high capacity).

Limitations:

- The possibility of 'channelling' which lowers interaction between the pollutant and carbon is vital to eliminate in GAC planned filter substitutions. Therefore, the efficiency and bacterial buildup in the filter are reduced.
- Often need filter changes.
- Can produce fines of carbon.

8.6. ACTIVATED ALUMINA

Activated alumina is composed principally of spherical beads of aluminium oxide (Al_2O_3), highly porous and have a huge surface area. The activated alumina surface area is 345–415 m^2/g. It does not shrink swell, soften or decay when immersed in water (Carr et al., 2012). Three types of activated alumina are used, including alumina sorbent, alumina desiccant and catalyst carrier activated. The granulated alumina has the internal active surface. A cartouche or container of activated alumina is circulated through tainted water. The alumina adsorbs the pollutant. Since t physical adsorption has specific limit, the activated alumina cartridge requires regular replacement.

Benefits

- Activated alumina can be customized by changing the procedure of activation and it will have a significant variation.

Limitations:

- This approach is not very capable of reducing the levels of other toxic pollutants. It requires additional assistance.

8.7. ZEOLITE

Zeolites are Si/Al-ratio aluminosilicates of 1-infinity. It has a tetrahedral network of silicon and oxygen atoms where oxygen is replaced by aluminium to produce aluminum-silicates in some silicon atoms. Zeolite is crystalline in nature character and can 's adsorb pollutants. The canals are widespread wax and hollows. Zeolites have 1–20 m^2/g in surface area. Sythetic zeolites is produced by hydrothermal processes at a temperature range of 90–100 °C and autoclaves are followed by ion exchange with particular cations (Na, Li, Ca_2, K, NH_4) (Rahman et al. 2012). The common characteristics of zeolites are good adsorbent, high exchange cation and molecular sieve,. It replace Na with the Ca_2/Mg_2 in water, the process of water purification is as follows:

Na Zeolite Mg_2 Ca=Mg Ca_2 Zeolite

Na: Natural zeolites are highly beneficial in the treatment of wastewater. Several natural zeolites (e..g., Clinoptilolites, mordenites, phillipsites, chabazites) have benefits of the selective division into NH_4 and transitional metal (e.g., Cu_2, Ag, Zn_2, Cd_2, and Hg_2) :

- Zeolite recharge is possible when the cations are exchanged with the original, and reuse also becomes possible.

Figure 8.5. Sodium zeolite.

Source: https://www.wateronline.com/doc/benefits-of-sodium-zeolite-for-drinking-water-treatment-0001.

- The removal of nitrates, phosphates, arsenates, fluoridated chromate and NH_4 and heavy metal from inorganic anions and of radionuclides (e.g., 137Cs, 90Sr, 60Co, 45Ca, 51Cr, 111mCd, 110mAg) is feasible.
- It can remove of organic and other humic (humic, fulvic and humin-including) compounds and odors.
- Capturing micro-organisms (the wide zeolite surface area is available to microorganisms for attachment). This results in the selection of an appropriate biofilter material to remove pathogens.
- Permeable reactive barriers (PRBs) in a waste disposal site may be prepared using zeolites to prevent contaminations from spreading to the ground water.

Limitations

- Tthe sewage sludge mass increases due insolubility and the usage of zeolites in the composition of detergent and softeners.

8.8. SILICA GEL

Silica gel is a strong glass amorphous grain or a silicone dioxide beaded substance (SiO_2). It is essentially a natural mineral that is refined and treated. Silica gel is a high-capacity surface adsorbent with small pores which can especially be employed as a desiccant, a moisture-resistant, rust inhibitor and catalyst.

It is generally silica gel is prepared using two routes: silica polymerizing and colloidal silica aggregation. Silicon acid, $Si(OH)_4$, is very polymerised, leaving minimum numbers of uncondensed Si-O–H groupings and forms a siloxane network (Si–O–Si). Van der Waals forces or cations bridge are formed during coagulation and then the aggregation takes place (Cook & Spray, 2012).. The first route of manufacturing silica commercially is by blends with a mineral acid, such as sulphuric or hydrochloric acid, a silicate solution for sodium. The reaction results in the concentrated dispersion of finely divided silica hydrosol or silica acid hydrated SiO_2 particulate matter:

$$(Na_2SiO_3 + 2HCl + H_2O \rightarrow 2NaCl + SiO_2 + H_2O)$$

The hydrosol polymerizes on its foot into a white, silica gel precipitate. The resulting gel is first washed, dried and then activated before use.

A range of silica gels having different surface area, pore volume and strength can be generated by changing the silica level, temperature, pH and activation temperature, (Iler 1979). The regular density gel has a 750–850 m^2/g surface area and an average pore diameter of 22–26 Å, compared to 300–350 m^2/g low-density gel and 100–150 m^2/g for the normal density gel. Silicium gel is utilized as a desiccant due to its huge pores volume and mesoporosity. The modified silica gel (model with an impregnation) with a high-molecular weight quaternary amine (triethyl octadecyl ammonium iodide) is used to remove heavy metals (cs, ag, hg, Cu, cs) from water (Dobriyal et al., 2012).

Benefits:

- Silica gel is a non-toxic and non-corrosive material.
- It has high adsorption capacity because of its very high surface area and porosity.

Limitations:

- Preparative aspects needed very precise control.
- Modification is needed to remove the specific contaminants.

8.9 ION-EXCHANGE

The coulombic attractive force between ions and charged functional groups is more commonly classified as ion exchange.

It is a typical reversible chemical reaction where an ion from a solution is exchanged for a similarly charged ion attached to an immobile solid particle. The selectivity coefficient controls the preference for ions of particular resins and is expressed as follows:

$$K_{B^+}^{A_+} = \frac{\{\overline{A}\}\{B^+\}}{\{A^+\}\{\overline{B}\}}$$

for the Exchange of $\overline{A}$ in solution for B^+ on the resin:

$$A^+ + \overline{B} \leftrightarrow B^+\overline{A}$$

Figure 8.6. Ion exchange.

Source: https://www.watereducation.org/aquapedia/water-treatment

In contrast with a solution phase, blocked phrases indicate the place on the resin (solid phase). The superscript and subscript on the selectivity factor show the reaction direction. Materials for the exchange of ions are insoluble substances that contain loosely held ions, able to interchange specific ions with ions in a solution passed by. Many of the natural substances such as proteins, cellulose, living cells and soil particles have the characteristics of ion exchange which are of significant importance for the function of nature. Dual shapes, e.g., beaded polymer matrix (resins), and membrane, can be created in synthetic ion-exchange polymers.

In addition to these things, the interests of various low-cost adsorbents are developing. Many agro-waste biomaterials, such rice-husks, soy shells, rice straw, sugarcane bagasse, tea leaves, palm-palm peculiar felt-sheaths, are discovered to be useful for this application. This is good for removing heavy metal ions at low concentrations (Pb, Ni, Cd, Zn, etc.). Biosorption is a quick event of the sequestration of nongrowing biomass by passive metal uptake (El Bedawy, 2014). Aspergillusniger, Penicillum hrysogenum, Bacillus firmus and Streptomyces sp. biomass also has the potential for the sequestrating of metal ions through solution forming metal complexes and for the retention of special growth supporting conditions. One of the new ideas of water treatment is diaphragm technology.

8.10. MEMBRANE WATER TREATMENT

In this case, a semi-permeable membrane is utilized to remove pollutants from water. Two types of water treatment membrane systems are available, notably pressure-driven (e.g., reverse osmosis) and electrically driven (electro-membrane) technology.

8.11. REVERSE OSMOSIS

The regulators of life are both processes (e.g., osmosis and reverse osmosis). Although simultaneously called concentration and pressure, both are governed by the system's thermodynamic function, i.e., 'chemical potential. The driving force is essentially described as the free energy change in the system as a consequence of the change in the structure of the system. Although the two technically mean the opposite process, they are similar thermodynamically. The trend for material transportation is always in a decreased chemical potential for both processes in the isothermal working condition. In osmosis, solvent flux occurs side by side across a semipermeable membrane, and the flux is a solution for solvent in reverse osmosis. Only solvent molecules migrate from one side to the other in both circumstances. The main two characteristics of a membrane process are flux and rejection. If an RO membrane is considered as permeating water only, the water and solute flux can be written as:

$$J_w = A(\Delta P - \Delta \Pi),$$

$$J_s = B\Delta c_s,$$

Figure 8.7. Expression of reverse osmosis.

Source: https://www.watereducation.org/aquapedia/water-treatment

where A is the permeability coefficient, and DP and DP are the hydraulic pressure and osmotic pressure difference calculated according to the following equation. The commercial membranes for ion exchange apart from the chemical structure (cation and anion) can be dividing into two principal categories, homogeneous, heterogeneous and heterogeneous, which may be further classified as monocardial (cation/anion) ion-exchange membranes depending on the degree of heterogeneity of ion exchange membranes (Fang et al., 2013). In terms of the chemical structure excellent selectivity and low resistivity, the ion-exchange membranes are very similar to typical ion-exchange resins. The mechanical requirements of the membrane process are mainly responsible for the difference between membranes and resins. Therefore, sheets of materials manufactured in the same way as a bead resin cannot normally be used. However, a membrane preparation with reinforcing support of a stable reinforced material provides the requisite resistance and dimensional stability is the most typical method.

Three different steps, that is, polymerization or polyconducting or monomers can be used to summarize the preparation method for ion exchange membranes. One must contain, at least, mobility which is anionic or cationic, respectively, introducing anionic and cationic moieties into the preformed solid film like a styrene-DVB based membrane and introducing anionic or cationic moieties.

Benefits:

- Non-toxic, security and reliability.
- Chemicals for regeneration can be completely removed.
- Effective to completely remove dissolved ionic particles, heavy metals, etc.
- Capability for treatment of greater quantities of SDI, TOC and silica feed water.

Limitations:

- It can only removal of ionic contaminaants withs low molecular weight.
- The technique cannot substantially remove uncharged, increased molecular weight and less moving ionic species.
- Wide membrane areas are necessary for low-level (and scarce conductive) feed solutions to meet capacity requirements.

8.12. ELECTRODIALYSIS MEMBRANE TREATMENT

The following three ways usually achieve catalytic processes: Nitrate hydrogenation, photocatalytic and electrocatalytic. Nitrate hydrogenation One of the potential ways for nitrate removal from water is the hydrogenation via catalytic approach. The reaction is preferably done at an ambient/low temperature. It demands very active catalysts.

Nitrate can only be reduced into the products such as NO_2, NO, N_2O and N_2. The undesirable NH_4 by-product also results from a hydrogenation side-reaction. Bimetallic catalysts (such as Pd/Cu, Pd/In, and Pd/Sn) have been supported as efficient nitrate hydrogenation catalysts. Other metals, in addition to Pd, (e.g., Cu, in, Sn, Co) play a promoting role in transforming NO_3 into NO_2 in the first phase. It is found that N_2 and ammonium (NH_4^+) are the stable end products of the catalytic reduction process. N_2, however, is regarded as a detrimental water contaminant(Zhao et al., 2017).

In photocatalytic approach, photographing speeds in the presence of catalysts of organic contaminants, infections, green algae or chemicals. As a reaction with UV light, the separation charges are activated and the surface adsorbed species are then scavenged and removed. TiO_2, ZnO, Fe_2O_3, heterogeneous photocatalysts have proven the efficient in the degradation of a wide spectrum of contaminants in water.

Photocatalytic processes are caused by UV lighting, which is equivalent to or more than the bandwidth of the semiconductor when absorbed by photoenergy. Therefore, the adsorbed pollutant species in water engage in a redox reaction. In addition to the process, the semiconductor oxidizes water to form OH•, a potent oxidant that interacts quickly with the contaminants in the water. Different techniques, such as inclusion of dopants, catalytic oxides of catalytic metal and mixed metal oxides, particle size and shape, are being used to boost catalytic activity using visible light.

In both, chemical and bacterial degradation, TiO_2 doped with nitrogen demonstrated outstanding photo catalytic activity compared to untreated TiO_2 nanoparticles.

Benefits:

- Catalyst reusability as the process unmodified.
- Reactions can occur at ambient temperature as well as low requirement of consumables.
- Simple operating process.
- Sufficient to treat low pollution concentrations.

8.13. CATALYTIC PROCESES

Catalysts with their fine particle size and huge area by volume ratio create a high catalytic clustering tendency while they are in operation, which means they are poisoned after water treatment.

8.14. HYDROGENATION

Catalysts with their huge area to volume ratio are crucial. The oxidation occurs via a surface mediator on anodised surfaces in electrocatalysis. The oxidization rate depends on the temperature, pH and diffusion rate of indirect electrolysis producing oxidants. The difference is considerably from electrolysis in which direct oxidation of pollutant occurs and oxidation rates rely on electrode activity, the rate of diffusion of contaminants and current density. For NO_3 removal, electrocatalytic reduction is widely utilized (Figuères et al., 2003). The creation of electrodes in this context is of interest.

Benefits:

- High degradation of pollutants, easy management and low cost.
- Easy to monitor with power up/off.
- Environmentally compatible because additional chemicals are little or not at all.
- It can remove various types of contaminants in bulk volume.
- The system runs in comparison to non-electrochemical processes at low temperature and pressure, therefore preventing the volatilization and discharge of unreacted waste.

Limitations:

- High operational costs due to high power usage.
- Fouling of electrodes can also happen on the electrode surface.
- The nature of the effluent should be known.
- The usage of metal ions can result in effluent of higher toxicity than the initial effluent, which may sometimes require an electrolyte addition. This technique, therefore, require separatio of metal species to retrieve them.

8.15. BIOREMEDIATION

8.15.1. Phytoremediation

It means that the usage of plants to eliminates contaminants from the environment. This technology involves multiple methods, such as phytoextraction, rhizofiltration and phytostabilization.

Phytoextraction entails the buildup of metal in the harvestable root and ground spring portions above. The absorption, planning and concentration of hazardous metals from polluted effluents are indicated by rhizofiltration. Phytostabilization is a method that reduces the mobility of heavy metals by tolerant plants while ohytotransformation/phytodegradation is a method in which pollutants can be transformed or degradaed by plant enzymes or enzyme co-factoring.

In the 1970s, the history of this particular study began with the absorption of harmful chemicals Hg, As and other metals. The macrophytes of water hyacinth (Eichhorniacrassipes), duckweed (Lemna minor), pennywort (Eichhorniacrassipes), water velvet (Azolla pinnato) are considered biological filters in this context and play a significant part in the maintaining of the aquatic ecosystem. The Lemna minor, Eichhorniacrassipes,Pistiastr atiotes and Salviniaherzogii floating plants have a good ability to directly accumulate the metals from industrial effluents (Fuhrmann, 2012).

8.15.2. Filtration of Bioactive Carbon

Another forward-looking bioremediation technology is biologically active carbon. The technique is based on the filtration of water using microbial biofilm attached to supporting material. The GAC particles can produce a 'biofilm' for a microbial (bacterial) colonisation. It is referred to as a 'porous tangled slime matrix mass.'

This is a microbial cell, either immobilised on a GAC (substratum) surface or incorporated into an external organic microbial polymer matrix. In biofilms, extracellular polymeric substances are secreted by bacteria and fungal cells into a cohesive stable matrix where cells are retained in dense concentration. The matrix is made up of polysaccharides, proteins, nucleic acids and lipids. Biofilm activity is concerned with physiological changes in certain genes or changes in the bacteria surrounding to improve local nutrient, oxygen and enzyme concentrations or to prevent the invasion of harmful or inhibitory chemicals. The majority of organic chemicals dissolved are physically removed in the GAC media through adsorption. Biodegradation can potentially work in addition to adsorption.

Advantages:

- Chemical disinfection treatment operations may be avoided.
- Due to organic substrate biodegradation on GAC medium, life span may be increased.

- Less potential bacterial regeneration.
- Eliminates the requirement for source filtration coagulant.

Limitations:

- The process growth is required to be controlled.

8.15.3. Photocatalytic Processes

The high-gradient magnetic separation (HGMS) is a frequent procedure in the magnetic separation process. This process involves an electromagnet consists of a magnetic bed of sensitive wires. Numerous parameters that are impacting the process are impurity kind, concentration, size, magnetic susceptibility, magnetic field strength, spacing design and magnetic field strength and its orientation. In general, three categories of magnet type separators exist, i.e., permanent magnets, electromagnets and magnets.

The permanent magnet (ferromagnets of iron, nickel, cobalt or rare earth) has magnetic fields less than 1T, however, it trends towards enhancing the strength of the magnetic field by developing characteristics of materials and form design. The electromagnetic device consists of a solenoid of electrically driven wires, when electrically driven through, may generate a magnetic field of 2–4 T in the cavity. The third magnetic separator category produces the maximum magnetic field intensity of 2 to 10 T (Vyas & Nath, 2021).

The magnetically supplemented water treatments may be classified primarily as the following type depending on the difference in the physical treatments such as direct purification, sowing and separation of organic and inorganic contaminants, including radionuclides, by magnetic flocculents and magnetic sorbents. No carrier magnetic component is used for separation in the direct purification approach. For purification, the fundamental features of the ions or the strong reaction to the magnetic field are used. This method usually uses the anti-scaling technique.

$CaCO_3$, $CaSO_4{\bullet}2H_2O$ and silica, $BaSO_4$, $SrSO_4$, $Ca_3(PO_4)_2$, and ferric and aluminum hydroxides are commonly used in magnetic anti-scale treatment. Coagulant cation forms an insoluble precipitate in the magnetic flocculant separation under the applied magnetic field. It is an efficient way of drastically reducing water effluent oil as well as suspended particles. Magnetic sorbents are not easily coagulable and wastewater purification is performed using ions (polymerizes as polyhydroxy complexes, or nitrous hydroxy or hydroxycarbonate or halogenohydroxycarbonate complexes) (van Koppen & Schreiner, 2014).

Benefits:

- Practicable for pollutant separation.
- Magnetic pre-treating improvises purification of RO membrane filters.

Limitations:

- Alone not enough for efficient water management.

8.15.4. Electrocatalytic Processes

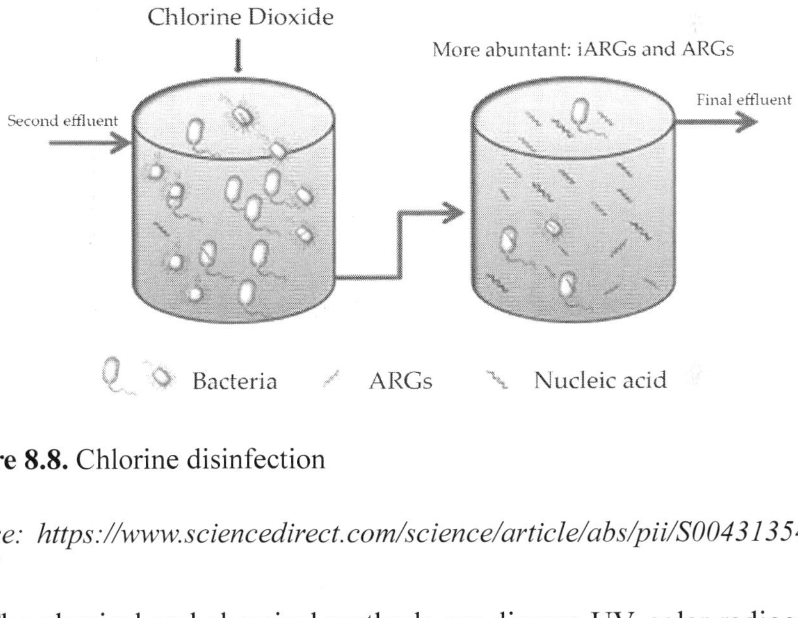

Figure 8.8. Chlorine disinfection

Source: https://www.sciencedirect.com/science/article/abs/pii/S004313541830 143X

The physical and chemical methods are diverse UV, solar radiography and ultrasonography are included in physical treatment, while chemical treatment involves chlorine, iodine and ozone. The therapy characteristics are described below.

Ultraviolet radiation, have germicidal effect where UV light from a low-pressure bulb travels through the water to be treated. The biological contaminants such as bacteria and virues are exposed to UV-light as the water passes the ultraviolet purifier and kill them by damaging theirgenetic components. Ultraviolet water treatment kills the bacteria. They are the pioneers of UV purification in water. One of the main drawbacks of this form of water treatment is that dissolved chemicals and other particle particles are not eliminated effectively.

Benefits:

- Easy to use and low cost.
- Good disinfection of germs and viruses.
- Do not require continuous attention.
- No influence of minerals in the water, excluding plastic bottles.
- 30NTU turbidity or less needed,
- Capabilities for the destruction or inactivity of a variety of pathogens in water.
- Ability to breakdown some organic pollutants.

Limitations:

- Toxicity can result from poor plastic bottle quality.
- Less active against bacterial spores and cyst stage of some parasites.
- Unfit for water with significant suspended particles, turbidity, color or soluble organic matter.
- Unable to work without electricity. Solar radiation is used as well. Pathogens, especially diarrhea, are particularly useful to inactivate. In transparent plastic bottles, the polluted water has to be filled and exposed to full sunshine for 6 h. The sunshine eliminates the virus with UV-A radiation (wavelength 320 to 400 nm).

Mechanical wave vibration can cause harm to bacterial cell structures. Therefore, disinfecting water can be advantageous. But it is also possible to regenerate the bacteria. The best method is therefore to combine it with the chemical disinfectant.

Benefits:

- Easy to use.
- No constant care required.

Limitations:

- Microorganism growth is also viable.
- Not completly self-sufficient.

8.15.5. Disinfection by ozoneç

Ozone (O_3) is a stable oxygen form and UV radiation protection layer in atmosphere. It is also an effective disinfectant for potable water. It easily releases oxygen and is hence a strong oxidant. Ozone is created using UV or a cold electric discharge by transferring oxygen. It can straightforward enter oxygen into chemical combinations to make up aldehydes and ketones with a very high oxidation potential of ozone.

Ozone oxidizes the organic components of the bacterial cell wall, thus weakening cell walls and leading to cell breakage. This exposes internal components of bacteria to environment, causing nearly rapid cell death. Ozone enhances clarity (clarifying iron, sulfur and manganese) of water. Soluble Fe(II) and Mn(II) are not being filtered in normal conditions but ozone therapy produce insoluble Fe(III) and Mn(VII) and therefore filtration is achievable. It also decreases odor problems and sulfur and other chemicals dissolved (Gao et al., 2014). The fundamental benefit of ozone is that it does not leave any residual disinfection in the water. It is created and administered immediately on-site to employ ozone as a disinfectant. Ozone is a major air pollutant, explosive, and irritant for the skin, eyes, respiratory tract and mucous membrane. The restrictions are severe. If there is little bromine in the water, it can develop carcinogens.

8.15.6. Disinfection with Chlorine

Figure 8.9. Excess chlorine and bleaching powder process

Source: https://aosts.com/municipal-wastewater-treatment-solutions/chlorination-dechlorination-system/

The chemical disinfection approach such as chlorination is effective in the destruction of biological pollutants (e.g., pathogens). Chloramine and chlorine oxide are the most prevalent and highly oxidized form of chlorine and its derivatives.

Chlorine is good for combating cyst-forming bacteria and protozoa (viz. Giardia lamblia and Cryptospordium). Chlorine gas management is risky, hence sodium and calcium hypochlorite are in the trend. Free chlorine is released into the water. Another way to achieve water chlorination is the electrolytic approach. When dissolved in water, free chlorine is released. The restriction of the application of chlorine in the water that combines with natural organic molecules generates potentially dangerous chemicals, such as trihalomethanes and haloacetic acids. It is shown that they cause cancer.

The maximum permitted average annual level is 80 and 60 lg/L respectively for trihalomethane and haloacetic acids. With pH, temperature, time and organic levels of water, they tend to rise. The reduction of organic substances is one technique of reducing the quantity of trihalomethane and haloacetic acides (Geissen et al., 2015). Therefore, removal of organic molecules from water is preferable after usage. In order to remove the organic products to manage trihalomethanes and trihaloacetic acids, EPA has advised to increase coagulation (the process of raising the feed rate of coagulants and adding stronger coagulants, perhaps ferric coagulants).

Chloramine is not a source of THMs or haloacetic acids, but a nitrification effect, as ammonia is a bacterial growth element with the production of nitrates as a result of products;

$$Cl_2 + H_2O \rightarrow HOCl + HCl,$$
$$HOCl \rightarrow HCl + [O].$$

Figure 8.8. Chlorine disinfection.

Source: https://www3.epa.gov/npdes/pubs/chlo.pdf

Benefits:

- Simple method.
- Availability of inexpensive chemicals/cheaply available.

Limitations:

- Excess of chlorine produces characteristic unpleasant taste and odor and irritating effect on the mucous membrane. However, the usual practice is bleaching powder in place of chlorine.
- Continuous provision of chemical materials and skilled personnel is needed for chlorine to be effective.
- The mechanistic process is as follows:

$$Ca(OCI)CI + H_2O \rightarrow Ca(OH)_2 + CI_2,$$
$$CI_2 + H_2O \rightarrow HOCI + HCI,$$
$$HOCI \rightarrow HCI + [O].$$

Figure 8.9. Excess chlorine and bleaching powder process.

Source: https://www3.epa.gov/npdes/pubs/chlo.pdf

Bleaching powder should only be used in calculated the amount because it gives a foul taste and an unpleasant odor, but less sterilizes the water.

Benefits:

- Simple procedure.
- Cheap chemical availability.
- It's a good oxidizing agent, like chlorine. In a very short period of time, it is effective against a wide range of pathogens, including spores, cysts, viruses, etc.

Limitations:

- Continuous provision of chemical materials and skilled personnel is needed for bleaching.

CHAPTER 9

THE ROLE OF GOVERNMENT AND THE PUBLIC IN THE MANAGEMENT OF WATER RESOURCE POLLUTION

CONTENTS

9.1. INTRODUCTION

When a person has access to a sufficient amount of water that is risk-free is normally a good indicator of human well-being. Water pollution is currently being considered as a biggest health risk to human beings and is known to be a global threat. This is so because a deterioration in the quality of water. Apart from the fact that it poses a danger to human beings, it is also a great threat to the entire ecosystem means other organisms are also at risk (Grizzetti et al., 2016). Scientists claim that the damage brought about by water pollution may be prove to be irreversible. This means that even though human beings decide to manage the same, they would still suffer the fates of the same for a very long time. Nevertheless, this should not be a reason for them quitting managing the water resources. It is high time for us to ensure the successful management process to make the world a better place for future generations.

Figure 9.1. Community taking part in the waste management process.

Source: CSIR I.

Cap-Net UNDP has taken steps to acquire knowledge of the best practices that will ensure that the water protection while at the same time, countries are trying to find solutions for preventing and managing the quality of water. There is unlimited knowledge regarding how to control pollution on the point, especially with centralized wastewater collection however, there is a limited knowledge about non-point waste management. This is about the wastewater that is passing through the rural and the urban settlements via the productive lands that support the communities as well. In addition to this, most of the wastewater the residential areas do not get treated due to lack of channelization and directly enter the water bodies. The pollution exerted on the freshwater bodies is growing at an immense rate.

Scientists recommend is that water pollution should not be considered as a technical problem alone but as a management challenge. This shows that great knowledge of the same is required to plan and implement the solutions to pollution problems and further strategize plans that will help in preventing and managing the same (Ghosh, 2017). Organizations are putting forth their minds and effort to educate the public on the best way to go forward about pollution management. Training materials have been provided to improve the practitioner's minds to understand water pollution and the stakeholders that need to be involved in the process to come up with the best strategic plans on how to manage and prevent water pollution. This may prove to be a difficult task but the stakeholders are the best representatives to ensure the success of strategies that are being adopted.

Figure 9.2. Stakeholders need to come up with better strategies to control waste such as this one.

Source: iPoint-system gmbh.

All stakeholders that are involved in the process of wastewater management need to ensure that they put maximum effort to achieve the progress and control water pollution. They should also consider the restoration of ecological balance to protect the world from the climatic and environmental changes. In order to achieve such objectives, both the government and the public need to work together in the process to ensure the success of the same. Apart from this, there is need for constant monitoring and evaluation by the national policies and legislation. Generally, the pollution levels are underreported due to a lack of monitoring facilities and frequency. The point source pollution which includes agricultural and industrial pollution are more visible than the non-point pollution (Hirji & Ibrekk, 2001). As such, individuals need to take an opportunity to work on the same since it is one thing that they can work with. Communities can work together towards finding solutions, which means that there will be a collective action toward pollution control. In order to efficiently address the problem of water pollution, there is need to for interdisciplinary action, where collective work from different disciplines will be done.

In order to ensure the success of the process, all the stakeholders need to be involved in the process. The polluters, t public that gets affected by the same and the government all need to take an initiative to work towards protecting the environment. In order to fully come up with better plans to prevent and manage water pollution, individuals need to fully understand water pollution and further able to understand the factors that are responsible for causing water pollution. In the world, there has been an increasing downward trend in the quality and quantity of water and many countries are experiencing dry rivers, accumulation of agrochemicals and heavy metals in fish that is normally consumed by human beings and other animals in the ecosystem. Apart from this, nutrients are lost in the water bodies as a result of pollution. The degradation of and overexploitation of the ecosystem are normally brought about by water users who are responsible for pollutant discharge. As per the current situation, immediate measures are needed to be taken to prevent water pollution and achieve water quality objectives. Two distinct actors need to work together to ensure protection and maintenance aquatic ecosystem. The two actors include the resource managers or regulators and the water users or what individuals call water polluters.

Figure 9.3. Water resource managers and users need to work together to control the water pollution.

Source: International Water Management Institution.

Minimizing the extent of pollutants and therefore protection of the country calls for a firm regulator, which is sponsored through clean guidelines and legislation and cooperation among the regulator and polluters, and a complete tracking and enforcement system. In addition, the different types of aquatic ecosystems such as freshwater and coastal water should be considered for maximum success. Water pollution control involves making plans through carefully thinking about ecofriendly approaches and wearing capacities of aquatic ecosystems.

The ecosystem is known to provide a wide range of services that contribute to water quality management. While management of water bodies especially the banks should be considered. The reason behind this fact is that the river basin is known to house a number of ecosystem aspects including multiple terrestrial and aquatic ecosystem, which include mountains, forests, grasslands, mangroves among others. In the context of water pollution control, green urban areas should be considered (Singh et al., 2020). Such an ecosystem, even though small, can complement the other ecosystem and rejuvenate the site polluted with different chemicals. When the ecosystem is healthy, it can absorb and degrade the pollutants such as sediments and excess nutrients. Wetlands are known to retain water and occasionally flood through the soil and then released into the rivers and groundwater aquifers. The presence of excess sediments are known to affect the health of the ecosystem, especially the rivers. Reeds and grasses are plants that are normally found in the wetlands and are known to increase the opportunity for sedimentation given the fact that it reduces the flow velocity of the materials.

Figure 9.4. Reeds and grasses in the wetlands help protect major water bodies from pollution.

Source: Enviropro.

Even though these sediments are known to destroy the ecosystem of the wetlands, but sediment holds the waste materials and avoid their entry the water bodies and maintaining the health of water bodies. As such, human beings should maintain the weeds and the grasses that occur in the wetlands as these plants would ultimately protect water bodies. In the current society, there is an increase in population and agricultural activities. The wetlands have been cleared for agricultural activities to fulfil the increasing demand of food.

The wetland is an important ecosystem and it possess certain plants, which play role in controlling. The plants of wetland decrease the mobility of pollutants and also degrade them to certain extend before entering the water bodies. (Jackson, 2006). If the plants get dried up, the pollutants can rapidly enter water bodies. This is the reason why prevention is always considered the best solution to problems that occur. Individuals need to understand the working system of the ecosystem as a broader aspect to manage the water resource.

Figure 9.5. Once the wetlands dry up, the waste can rapidly move to the major water bodies hence wetlands are important for maintaining water bodies.

Source: ResearchGate.

As a way to reduce or rather manage the pollution of the water resource, human beings should be careful about their activities. Human activities are also part of ecosystem and as such their participation in the management and controlling the water pollution is also important to conserve the ecosystem. There are some things that individuals need to consider to ensure that they become stewards in the process of ensuring that they are protecting the environment.

One thing is that individuals should not dispose the oil and grease from cooking or any other activity directly to drain, it can be collected as oil is immiscible in water. Fat is lighter than water and tends to float on the water.

In the process of floating, it blocks air circulation in the water bodies and may kills the organisms that reside in the particular ecosystem (Kim et al., 2015). Some organisms are important in maintaining the water ecosystem while other act as edible food. Taking for instance fish, which is not able to survive under lack of oxygen in a water body. This will affect the food production rate and the demand for food will not be fulfilled.

One should be careful not to discard any pills or any forms of chemicals to the sinks and the toilets. There are proper recommendations regarding disposal of these products. The reason behind this fact is that, the pills may be containing some harmful chemicals that are harmful to the ecosystem in the water. Others may contain heavy metals that when ingested by the fish accumulate in their bodies. The same may be harmful to other humans given the fact that pills are normally prescribed for an individual for a particular illness.

Figure 9.6. Individuals should not dispose of pills and other chemicals in the sink or toilets as it goes to the water bodies.

Source: Alamy.

Human beings should also ensure not to dispose the cleaning agents in the sink or even flush them in the toilets. In such cases, an individual can find the agencies that are responsible for collecting these materials. This will help prevent the pollution problems provided these agencies have proper knowledge of disposal of the same or whether one needs to recycle the same.

Another aspect is that individuals need to learn not to use the toilet as a waste bucket. Paper goods including tissues, wrappers and even dust clothes should not be discarded in the toilet but rather in the waste bucket. Initially, it was quite easier given the fact that people were known to utilize plats rather than tissues unlike in the current society. Unlike leaves, waste papers do not easily decompose proving to be quite difficult to take care of because these materials are fiber-reinforced. It is very popular in the current world and will not be left any time soon (Krampe, 2017).

Individuals should stop putting all waste into garbage. The solid recyclable waste materials can be collected and later recycled with help of the waste managers. In the case of wet waste, such as vegetable craps, fruit waste, leftover food etc. can be collected in form of the compost pile and used manure for farming. Manure prepare in such a way possess no chemicals and ecofriendly as compared to the synthetic fertilizers.

The dishwasher and clothes washer should be used when one has a full load. Human beings tend to use these machines even when they do not have a full load due to convenience and zero labor. Using these machine with appropriate load will save electricity and water at the same time. Apart from that, one should not use excess amount of soap or detergent. Also, the use of phosphate-free soap should be encouraged.

Figure 9.7. It is important to load the dishwasher once there is a full load. The same case also applies to the clothes washer.

Source: Stacy Risenmay.

Individuals should use a water-efficient toilet, which use optimum water per flush and avoid wastage of water. One should also consider the quantities of water available for use in order to preserve the water from pollution. Home with cellar drain, should ensure that the water does not lead to the sanitary sewer system. Also, make sure that the untreated system does not mix with the treated one if happens, should be fixed immediately (Kirschke & Newig, 2021).

Last but not least is that individuals need to be careful about the pesticides and fertilizers that they use in their farms and gardens. If possible, the best practices involve the use of natural methods in the farms. That includes using natural manure and natural pest control mechanisms. If one cannot do so, then the chemicals should be used at minimal levels. The containers of fertilizers and pesticides should not be disposed of in the sanitary sewers rather than use an appropriate disposal method should be followed.

9.2. THE ROLE OF THE GOVERNMENT IN MANAGING WATER RESOURCE POLLUTION

Government plays a vital role in the management of water resources. They have to ensure that the people under their jurisdiction have access to clean and safe water. Water as vitality to the existence of human beings and it need to be protected. Water is essential for the smooth working of different organizations and almost every company on earth. People of developing countries do not have access clean drinking water and their sanitation needs are hardly met. The most common types of negligence by the government include; excessive withdrawals from river flows, pollution from industrial and agricultural activities especially from big farms, nutrient loading which eventually causes eutrophication, the infestation of water resources by plants and animals, excessive fish harvesting, alteration of habitats, change in the flow of water and sediments and inadequate or inappropriate and poor infrastructure.

It shows a failure and fault in the management, planning and decision making from the managers of the area if the water need of a population or residents cannot be met from the available water resource. The planning, management and development of water resources should be done very efficiently to ensure the quality, sustainability, adequate and inexpensive water supply to both humans and the natural environment. This is very much achievable with the help of individuals organize, and addressing the factors such as illiteracy, population, poverty and corruption (Larson & Lach, 2008).

In the last few centuries, surface and groundwater has been a source of water for industrial, agricultural and municipal consumers. Sources such as rivers provide water that can be used for the production of hydroelectric power and large rivers such as the Mekong River are used for it.

Figure 9.8. The Mekong river is an important transport corridor in Asia.

Source: Heifer international.

Transportation of bulk: The rivers can also be a source for recreational activities and water for drinking by wildlife. The quality and quantity of water from rivers will determine the health and diversity of the riparian land and ecosystem.

The individuals responsible for the different water resources either regional or national should manage them properly to prevent pollution and prevent over usage of the resource to prevent depletion and mismanagement. There are serious consequences of water mismanagement and may result in pollution of water and later water depletion. Misuse of water is mostly brought about by the imbalance between requirement and/or demand and what nature or the water resource may offer and/or supply.

Governments should urge their citizen to manage, monitor, treat and supply water especially in urban areas, this has to be done efficiently with the use of important measures (Singh et al., 2021).

The most basic role of the government is to supply clean, quality water from the three main water resources and transport it to its citizens. Governments should supply water to their citizens and it involves three main processes. 1) Procuring; governments obtain water from ground or surface water and in some instances have to harvest rainwater. The procurement of water is accomplished by facilities that withdraw water from reservoirs, lakes and rivers. The government should oversee this process and ensure good integrity of the water source is maintained. After the water has been procured it has to be treated, 2) treatment; the institutions that offering water treatment facility are tasked with water treatment process up to required quality. After treatment, the water is delivered, 3) delivery; the water then has to be delivered to the consumers. A survey conducted in the united states showed that up to 90% of people depend on water delivered to them by their respective governments. The government has to maintain and monitor the different infrastructure that is used for the supply and storage of water. There are monitoring agencies that are independent of the water supplier that ensure that water is clean, oversee treatment facilities, i.e., ensure chemicals used for water treatment is harmless to the consumers. These institutions monitor the quality of treated water by measuring mineral content, water purity and any changes to the water supplied (Sroufe, 2018).

Figure 9.9. A water treatment facility.

Source: IndiaMART.

The supply of water is very crucial and with the government being obligated to carry out the process, if individuals were to look at water by themselves it would be chaotic but with centralized facility, it becomes easy to manage. Individuals pay their water bills based on the rates decided by the respective authorities which may vary from place to. Most of the individuals pay in terms of the amount of cubic meters of water used. The money collected is used to maintain the whole water supply system that include payment to the workers involved in the water supply and maintenance carried out on the supply infrastructure, treatment, water quality monitoring etc.

The government also funds water projects in their respective authorities. The funds come in form of budgetary allocations, grants and allocations which are used for different purposes about water. Most of the funding from the government comes from taxes submitted by the citizens. The money that is obtained is used for different purposes.

The institutions obligated with the supply of water are also obligated with ensuring the waste that comes from home has to be treated before it is released into other waterways. In countries such as Kenya, the water and sewerage companies are run by local country government. They ensure water reaches homes and they also regulate water supply. The waste from homes are directed to sewerage areas where there are large tanks for the sewage storage and also treatment before release. The wastewater from bathrooms, toilets, kitchens and laundry rooms are redirected into a single pipe and transported to the sewage treatment plants. The solid wastes are separated from the liquid and the government sometimes converts the fecal matter into fertilizer or sometimes used as a fuel. After treatment, the liquid is safe to be released into the natural environment as all the chemicals have been removed or neutralized.

Figure 9.10. A sewage treatment plant. Sewage is treated to possible extend and then released.

Source: international water association.

The government funds new projects and also awards grants to researchers and students to look for other ways of acquiring water apart from relying on groundwater, surface water or rainwater. These sources will not always be enough for the consumers because of the fast growth rate of the human population.

In order to manage water resources properly governments whether local, regional or national have to come up with policies that are cross-sectoral with respect to the natural environment, development, energy, economy, trade and agriculture to will improve the integrated management of water resources and its governance. For instance, the governments need to have agricultural policies that are beneficial to the protection of water resources (Leahy & Anderson, 2008).

Governments need to engage people to establish mechanisms that will encourage companies and different industries to be efficient in the use of water resource and be transparent in how they use water. This can include the creation of legislation and/or agreements voluntary or involuntary that will promote transparency and disclosure on water usage. The government can also include citizens to raise awareness on how to consume and reduce water usage. Across the boundaries, the government can increase foreign aid to promote sustainable water management of water resources that need protection. The different governments can ensure that in the long run freshwater resources are managed sustainably of their.

Figure 9.11. Government officials involving citizens in decision-making.

Source: gov.uk.

Set the maximum sustainable limits of water that is consumed. The limits should also be set on the pollution of water in rivers, river basins and aquifers to ensure the balance between water that people are using and what nature needs and can supply.

Establish equitable sharing of water, i.e., sharing water among people equitably in that all the different social classes get access to water regardless of where they live. Ensure there is fair sharing of water footprints within the rivers and water needed by the natural environment, i.e., consuming water without disturbing the natural ecosystem around the water source and amongst the people.

The government should always be updating their statistics, in the past only available statistic is the amount of water withdrawn from the different waterways. The statistics have been generalized hence it may sometimes become quite difficult to manage the different types of consumers (Lopez et al., 2009). The government has to come up with statistics to show the domestic consumption, industrial water use and water that has been used for the production of exported items. Government can easily manage the statistics by using available manpower and resources and can take further steps based on statistics to efficiently manage water consumption. This will also help to understand the type of wastewater each consumer is releasing into the environment.

The local government needs to set the water quality control and acceptable parameters. For instance, in Japan water monitoring is usually conducted every four times per day. This means that water samples are collected every six hours and tested for pollutants and any foreign material. The changes in water quality are monitored to observe the effects of different pollutants on the environment. Lakes are also monitored but it is carried out seasonally because lake water quality do not change often. The pollution control measure carried out or undertaken by the local government include;

Figure 9.12. Collection of samples. Japanese authorities collect samples every 6 hours.

Source: ALS environmental.

Regulating factories using set laws and ordinances. The legislative arm of government has set laws that are required to be followed by the different industries as soon as they set up and in their day-to-day running. The factories for instance are required to treat their wastewater before it is released into the environment and the smoke from the processing zones has to be purified before release. Industries acquire the water from a recommended source. All the waste from the factory either organic, inorganic or any kind of trash has to be disposed off carefully away from waterways or water points.

Maintenance and construction of new sewerage lines and sewer treatment systems. The local government should maintaining the old sewer lines to ensure they do not leak and release the waste into the environment. Due to the expansion of towns and cities, the local governments are setting up new sewer lines and also they have city plans that show where sewer lines are supposed to pass hence no construction on top of sewer lines (Prinz & Singh, 2000).

Purification of water in waterways. Due to the constant use of waterways by different machines such as ships, motorboats etc. It is require that the water is purified to ensure that the water remains harmless and do not to affect the living organisms inside the waterway.

Aeration of the Sagami Lake. Lake Sagami in Japan was created in 1947 after the dam was built of Sagami River. The main purpose of water use was recreation and production of hydroelectric power. It was also used in the Tokyo summer Olympics for canoeing. Water to the lake is supplied by the Sagami River. By the year 1984, there were almost 170 different companies and around four sewage treatment plants that were surrounding the lake. The government instructed companies to treat waste before releasing into the lake. The lake has been experiencing eutrophication issues and this was observed in 1967 when there was a large algal bloom. This means the algae have to be removed to improve the oxygen concentration in water.

Figure 9.13. Lake Sagami in Japan. It is usually aerated because of eutrophication.

Source: TOKYO WESTSIDE travel guide.

Control of pollution from the new high tech industries, which are producing electronic products. These new industries are producing high tech gadgets, which need capacitors and other types of chips to be assembled to make an electronic machine. These gadgets use ultrapure water. These industries use a lot of heavy metals such as lead, chromium and others which are very harmful. The government requires monitor the wastewater that is released from these industries to ensure they do not release harmful wastewater into the natural waterways and reservoirs. The government has set guidelines, rules and regulations which have to be followed and in case of any overlooking these regulation by companies, appropriate action should be taken by government.

Many countries have their different laws but they all aim to exert a positive influence on the regulation of water pollution. In Europe where most countries are under the European Union, share similar laws and their main aim is to protect the ocean surrounding Europe and prevent marine dumping.

The dumping of industrial waste was terminated in all countries in the northwestern part in the year 1989 and the United Kingdom terminated waste dumping into ocean in 1993. There was also the termination of dumping of sewage sludge in 1990 and the United Kingdom also joined later in 1998. They have also set standards to be followed for quality drinking water to ensure safe drinking water for their citizens.

Figure 9.14. A parliament in session. They come up with rules and legislation.

Source: Minnpost.

The main role of governments is to come up with legislation that will help to combat water pollution. The legislation should be practiced strictly to bring any positive change. Most countries come up with the best legislations but the negative factor comes in while implementing it as in most cases the governments do not focus on implementation or sometimes partially implement them. In most instances, the government individuals are there to serve their own interest hence they end up sacrificing the citizens as they believe that people will come and go and the opportunity to make money comes rarely hence they end up taking advantage of the opportunity they get.

There are many laws, guidelines, rules, regulations, orders, acts, and ordinances that have all been set by the governments but the different cooperation's are taking advantage and lining the pocket of politicians to make their ways. If all the government rules arefollowed responsibly, then there would be less pollution and in most instances, the citizens would have safe, pure and enough drinking water for consumption in their homes (Partha et al., 2019).

If all governments do away with corruption and bring in humanity to ensure they do not serve their own interests when in power, the legislation will pass and be implemented and everyone will benefit most especially the future generation. The government's major role is to fund the different activities. When they fund the different departments well then they will perform their functions properly.

CHAPTER 10

BOARDS TACKLING GLOBAL WATER POLLUTION

CONTENTS

10.1. INTRODUCTION

It becomes more expensive and difficult to remove certain pollutants from water. Till date, more than 80% of the global water remains untreated which shows in the level of trouble in accessing good quality water. Pollutants from industrial water, farm chemicals and human waste usually mix with water. It is important to understand the effects of water pollution on the availability of water for humans use like bathing, drinking, cooking among other things. In addition to this, it is quite clear that the pollution of freshwater normally affects the quality of the ecosystem life that uses water as their habitat (Ludwig et al., 2014). This includes fish among other organisms. The freshwater is unfit for use when pathogens are present which is normally com from human or animal waste, organic matter used in the farm such as herbicides and pesticides, chemical pollution from toxic industries and salinity by runoff from mines and rivers, domestic wastewater and irrigation.

Figure 10.1. Water for cooking. Water is an important aspect of the life of human beings as it is used for drinking, cooking, bathing, agriculture among other things.

Source: https://vnifood.com/trick-for-adding-water-to-the-stir-fry/

After the industrial revolution, there was increased use of plastic for various purposes such as carrying material like food, medicine etc. The problem is that these things are currently polluting our environments because of poor disposal. Due to water pollution, the populations around the world are currently facing a major water crisis and are forced to pay for the little resource that is remaining. This has captured the attention of the authorities across the globe. Since this is a global matter, all the countries should come together to tackle this issue. There are a number of boards that are working foot and mouth to deal with this global menace. One of them is the United Nations Environment Program (UNEP).

10.2 UNITED NATIONS ENVIRONMENT PROGRAM (UNEP)

UNEP is the leading authority that deals with all the environmental related issues. UNEP sets the global environmental agenda and further promote the implementation of environmental dimension and policies that are supposed to bring better results for environment protection. The board is developed under the umbrella of the United Nations (Muller, 2015). Ideally, this board works for environment at world level. The mission of UNEP is to provide leadership and further encourage the partnerships that beneficial for the environment through the process of enabling, encouraging, informing and further inspiring individuals around the world for improving the quality of life by taking care of the environment and not compromising the lives of the future generations.

Figure 10.2. Logo of UNEP. UNEP is founded in 1972 and will celebrate 50 years in 2022.

Source: Green Growth Knowledge.

UNEP has set shops in all parts of the world. Its headquarter in Nairobi, Kenya works through their divisions, their regional and out posted offices and works with the various centers of excellence. In addition to this, they normally host several meetings regarding environmental conventions and is led by an executive director.

UNEP normally categorizes their activities into seven broad thematic areas, which include disaster and conflicts, climate change, environmental governance, ecosystem management, resource efficiency, environment under review, resource efficiency and chemicals and waste. This shows that pollution is considered as a vital component. This is known to be affecting the quality of human life. Pollution is has resulted in increased mortality rates especially in the countries that have revolutionized more than others have. Taking the case of China where industrialization is on another level. The quality of life is good only for those that have the capability of affording better quality water but this is not the case for those who re living below the poverty line (Nazemi & Wheater, 2015).

The work of UNEP is normally made possible by various investors. 95% of their income comes from voluntary contributions from individuals across the globe. UNEP appreciate the individuals for their contribution by honoring them every year. It hold various conventions that involve research bodies to bring critical multilateral environmental agreements to for nations and so that the environmental community at large can handle some of the greatest challenges of time. The Governing Council of the United Nations Environment Programme,1 at its 27th and primary usual consultation held in February 2013, in paragraph 17 of selection 27/2, determined to decorate transparency and openness in its work, and on this regard asked the Executive Director to set up a written access-to-facts policy (Owusu et al., 2016).

Figure 10.3. The governing council of UNEP. They assemble to formulate environmental policies.

Source: The Access Initiative.

In addition to this, the Executive Director mounted the UNEP2 Access-to-Information Policy on 6 June 2014, which was applied on a period in-between basis. A technique of its review, the Executive Director established the existing Policy that superseded the above-stated period in-between policy. The Executive Director acknowledges the order for UNEP to absolutely perform its mandate entrusted to it with the aid of using the General Assembly in its decision 2997 (XXVII) and different applicable resolutions. it's miles is considered to be significantly crucial to disseminate and make the records regarding its work or records generated through its program, especially with the recognition to environmental records as broadly as possible. In a precise sense, within the context of Principle 10 of the Rio Declaration on Environment and Development, it's miles of essential significance for UNEP to make environmental records to be handed to the Member States, different stakeholders and the general public preferred as a way to assisting their efforts in attaining sustainable development goals (Olsson & Andersson, 2006).

With regards to the above-stated policy, the information concerning the environment whether written, electronic, visual or urial are vital and are formed based on several aspects. The first is that, the elements that are in the environment which include, water, soil, air that we breathe, the land, landscape natural sites, biological diversity and the components and include the ones that are genetically modified and the interaction of these elements fall under the UNEP body and are crucial when considering the implementation of a policy. The aspects that are within the environment such as radiation, energy and noise, measures and activities and other aspects that have the likelihood to affect the various elements within the environment fall under the aspect of the first issue discussed. In addition, economic analyses including the cost-benefit analysis and the assumptions are vital importance and need to be taken into consideration when making decisions regarding the environment. Another aspect that falls within the jurisdiction of UNEP is the human safety and their health and the conditions that make up the life of human beings, such as their cultural sites and built structures. As much as the elements are affected by the state of the environment, are important aspects that also interchangeably affect the environment and are of great consideration when making environmental decisions.

Figure 10.4. UNEP plays an important role for the public. Is a body within the United Nations that is tasked with protecting the environment.

Source: UNEP.

Furthermore, it is the duty of this body to handle all the sensitive information regarding human life and the environment in an appropriate manner and maintain consistency with the relevant rules and regulations of the United Nations. The endeavors regarding this policy play a major role in ensuring that a balance is a stroke between the need to grant the public with the relevant knowledge regarding the UNEP's possession which includes the environmental information and the fact that individuals around the globe need to respect and provide confidence to the information that is provided (Olsson & Andersson, 2006). The executive in this case should review and update the policy as required. In addition to this, the policy will be considering the inputs of the member and all the relevant stakeholders. This is to ensure that the policies that are going to be implemented have better coverage in terms of the advantages that it provides to most if not all countries. The target of the body is to implements policies that can benefit all the countries especially the member states.

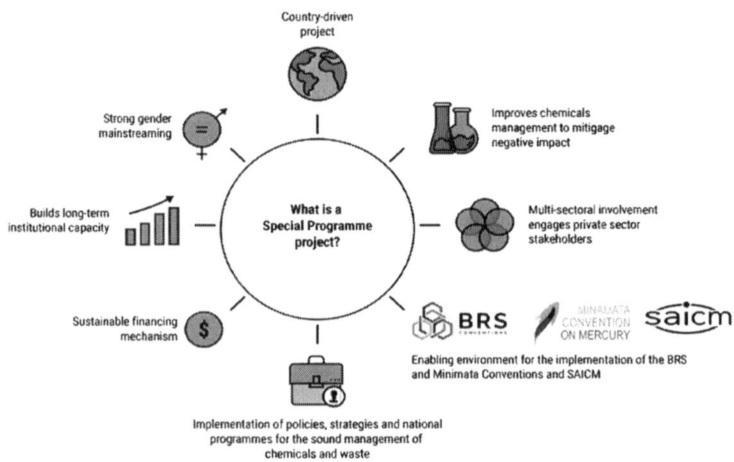

Figure 10.5. UNEP policy implementation. The process of implementing policies by the UNEP.

Source: UNEP.

The commitment of UNEP is to make the precise and required information that is related to the work that is tasked with, in particular information regarding the environment and the public at large. The policy formulation is normally guided by the fact that the organization consider the openness of the individuals that are involved. Any information that falls within the lap of the body, in particular, that which concerning the work of the body should be in accordance with the mandate that has been provided by the general assembly in resolution 2997 (XXVII) and other relevant resolutions (such as resolutions 47/190, S-19/2 and 66/288). This includes all the information regarding the environment that are normally formed and maintained by the UNEP and is normally available to the public (Olsson & Andersson, 2006). The information regarding the functions and responsibilities of UNEP available to the public are:

- Promotion of policy guidance with regard to the directing and coordinating the environment as is recommended by the United Nations system.
- Promotion of international cooperation with regards to the field of environment and recommend appropriate policies that will ensure that the environment is better maintained.
- Ensuring that the condition of the environment around the globe is kept under review and the environment associated emerging problems are considered as significant issues and provided great attention with their respective governments.
- Ensuring that the periodic reports provided by the executive director are reviewed and policies regarding the environment programs by the United Nations have been implemented.
- Promoting the contributions provided by other agencies across the world regarding environmental knowledge and information that proves to be appropriate to the technical aspects of the environment.
- The national and international policies regarding the environment are continually kept under review especially when dealing with measures that affect the developing countries especially alter additional costs that are incurred by these countries in the process of implementing the environmental policies. Most developing countries do not have the resources to implement some important environmental aspects and need a boost. Furthermore, they need to be ensured that the policies should be implemented to achieve developmental goals of the country.
- Approving the program for utilizing the resources that fall within the Environmental Fund.

Figure 10.6. Water pollution management and control is one of the objective of UNEP. This body is tasked to maintain the quality and enough quantity of water for the users.

Source: NRDC.

UNEP bear many responsibilities with regards to pollution management and further ensure that the control of the same has been made inevitable. In 2016, the body conducted a preliminary assessment of the water quality of the rivers at Latin America, Africa and Asia and the results were not at all pleasing. A third of the rivers was found to be severely polluted with pathogen and a seventh of the rivers have been affected by organic pollution while a tenth of the rivers in the regions that were under review had moderate salinity. The role of UNEP is to ensure that the environment is kept safe. One aspect is that the body was involved in The Global Program of Action for the Protection of the Marine Environment from Land-based Activity. This activity was known to concentrate on the reduction and the regulation of wastewater, marine water litter and nutrient loading (Olmstead, 2014).

This water program supported countries around the globe to monitor and generate water quality reports in their regions. The body even went further to provide many solutions that will ensure the management of water resources which includes water quality maintenance. In addition, the body is global custodians of the sustainable development goal 6.3.2, which is recognized to assist countries to fully understand, measure and report the ambient quality of water. The fact that water pollution has been seen to be a menace around the world, it can work with other bodies in ensuring that the water quality is maintained. Another body that is involved in wastewater management is the World Health Organization (WHO).

10.3. WORLD HEALTH ORGANIZATION

Figure 10.7. WHO also takes part in the formulation of policies regarding water pollution management. WHO conducts seminars and conferences for this purpose.

Source: NPR.

There has been an increase in concerns about water pollution around the world. Therefore, all countries around the world are working tooth and nail to achieve sustainable usage of water resource. It has been agreed by countries around the world that there needs for better policy framework that guides individuals on how to maintain the quality of water resource. There are a number of elements that have been identified and have been known to be helpful in the protection of water resource. Within the legislative frameworks of most countries, there is an aspect involving pollution control in most cases, the statements are normally hidden in official documents. They are normally included in the plans and actions that need to be done by the government. Most constitutional documents and government statutes include clauses on environmental policies for environment protection and conservation. Most of these statements are normally not straightforward and in most instances are inconsistent with other policies. Since they are developed separately with different purposes (Olsson & Andersson, 2006). In normal situations, water pollution is normally addressed in connection with the establishments in environmental regulations and action plans and are normally within the framework of water resource planning and management. Ideally, the documents involving the health welfare of human beings includes segments of water pollution management. Water pollution is considered as a very strong health hazard, therefore, the organizations that are involved in well-being of human health tend to consider the measures related to water pollution mitigation. Therefore, WHO also takes initiatives regarding water pollution management.

Figure 10.8. WHO ensures the health of the public is maintained. It looks at the well-being of the public and ensures food security and nutrition.

Source: NCD Alliance.

Typically, three interacting areas normally work together to ensure water pollution control. They include the ministry of environment, the ministry of water and the ministry of health. When the policy implementation process is carried out, these bodies work independently. For the success of any government policy, there needs for on-ground implementation of work. As such, it is very important for the countries to clearly state the policy statement and further provide proper policy documents that will act as guidelines in wastewater management. The recommendation is that the policies that involve water management need to be included in the water resource policy document or the environmental policy document. In the policymaking process, there are general principles that need to be put into consideration.

Firstly, the water pollution control policy needs to be considered as part of a coherent policy. They should be originated from the overall statements that are normally found in the constitution and government statutes to specific policy statements that are normally stated in the water or environment resource management. The second thing is that the policymaking process should include consultations from expert from the relevant field. Organizations that are involved in economic development policies should also be included in the process and in the instances wherein policies for other sectors are made; the water resource management should be put into considerations. This is aspect very important for the existence of human life but generally, disregarded while making policies. It should rather form part of the policy formulation process. The third aspect is about the policy statements regarding water quality management, that need to be realistic and easy to follow (Olsson & Andersson, 2006). Taking for instance a statement such as "No pollution of surface water shall occur..." this cannot be applied in practice even though it has good intentions given the fact that achieving such will prove to be quite difficult. As such, if such is stated, it will be seen as something that can be meaningless in the class of operational policy. It is important to consider the policy statements in the process of formulation, as they need to be ones that can be easily achieved. The fourth aspect is that the policy statements need to be long-lived. This means that when formulating policies, the body should consider the fact that they need to be able to pass a laborious political implementation process. So there should be guidelines that will ensure a regular adaptation of the same and synced to the country's developmental goals. In addition, guidelines should be included within the regulatory framework, which will ensure amendment of the policies on short notice.

Figure 10.9. Guideline on wastewater management. The guidelines help individuals understand the need to maintain the environment.

Source: ResearchGate.

A policy file has to be formulated virtually and concisely, and at the same time, it needs to be operational. The statements should be understandable without problems, the file ought to shape a manual for directors formulating legal guidelines and policies as properly as the ones enforcing, and thereby interpreting, such texts. To fulfil those requirements the policy file needs to include, further to very fashionable statements, properly defined guiding standards for water pollutants control in addition to outlines techniques for the implementation of the coverage. In conjunction with other bodies, the WHO has come up with a number of guiding principles for controlling water pollution. First, there is a need to understand the WHO.

The World Health Organization is a specialized agency that falls under the umbrella of the United Nations. It is tasked with the responsibility of public health. The board contains a constitution that establishes the governing structure and the main objective is the attainment by all individuals the highest possible level of health. By virtue, this is the greatest thing achievable by the world. The headquarter of the same is located in Geneva Switzerland and has six semiautonomous regional offices around the world. It was established on the 7th day of April in 1948 and is up to date commemorated as world health day. The broad mandate of this body is the advocating health care, monitoring world health risk, promotion of human health and wellbeing and monitoring all risks associated with the health of the public. It is known to provide technical assistance to countries especially those with limited resources such as the developing ones. The main role of this body is to set international health standards and health guidelines and further conducts surveys on global health issues (Olsson & Andersson, 2006).

The body is known for its flagship publication, which is the World Health Report, which provides a better assessment of the world health reports and global health topics. In this report, one would be able to see a clear picture of world health. It further provides a forum for summits and discussions regarding global health issues. The WHO has led a number of public health achievements which include but not limited to the near eradication of polio, eradication of small pox and the development of the Ebola vaccine. There are diseases such as HIV/AIDS that fall under the category of diseases that are currently given much priority. In order to ensure the success of the agency, it relies on the contributions from the member states. The states that are endowed with resources are normally contribute while the developing nations depend wholly on the World Health Organization for assistance regarding health matters. Some private donors have taken the initiative to provide charity to this organization. Currently, the approved budget for this body is over 7.2 million dollars. The majority of the contributions come from the member states. The major contributors include Germany and Bill and Melinda gates who contribute 12.8% and 11.65% respectively and the

United States, which is contributing 7.85% of the budget (Zhao et al., 2017). The fact that WHO deals with the health of the public, it has taken initiatives to come up with principles in conjunction with other boards to deal with water pollution. The general principles are discussed below.

Prevent pollution rather than treating the symptoms of pollution. When one looks around the world, one can see that pollution can damage the world. Nevertheless, one should not lose hopes, as pollution can be reverted. With past experiences, one can easily tell that the remedial actions have only gone along the way. These actions include the cleaning of pollution sites and in conclusion, the process has proven to be much more expensive than controlling. Wastewater treatment equipment has been installed and have been improving all through the years but water pollution still remains a big problem even after all these years. Sometimes, the introduction of a new treatment plant made the situation worse than it could have been. This is because of the formation of substances such as wastewater sludge. As such, the most logical approach that seems to be working at the moment is the prevention of pollution. This involves the minimization of wastewater.

The second principle is the use of the precautionary measures. There are many cases in which human beings are seen to release hazardous substances to the aquatic ecosystem even when it is clear that such substances would cause detrimental effects to the environment. One would just say that human beings are careless and do not considerate importance of the environment. Up to date, the release of such substances o the aquatic ecosystem is accepted widely and unless there is scientific research that clearly link the cause and effects of such substances to the environment, there is no single person that will be willing to change their ways. Even when such evidence will be provided, it might prove quite late given the fact that damage to the environment has already happened.

The creator of pollution should pay is the third one which means that the cost of the pollution, the control, prevention and reduction measures should be paid by the individuals that are responsible for it. This is not a new concept but has not been implemented and it can put individuals in check. Around the globe, it is recognized that water is a free commodity and individuals tend to use it as it pleases them including the fact that they pollute it at the same time (Afroz et al., 2014). This principle is considered as an economic instrument that would be able to help individuals to behave appropriately. It agrees to the fact that, by inducing a particular behavior, there will be less strain put on the environment. One example of such aspect of reducing wastewater discharges is on the financial charges that are sked on the industries that usually discharge wastewater. This has been a motivation for industries to behave well when it comes to the disposal of their waste.

Figure 10.10. Industries pollute water. Polluter pays principle may be the best principle yet.

Source: Water Integrity Network.

The other principle is applying realistic standards and regulations. This is considered as an important element in the implementation of a water pollution control strategy. The standards and regulations need to be achievable and enforceable. They need to be tailored in a manner that will match the level of the economic and administrative capability and capacity. To ensure success, the standards need to be regularly tightened to achieve the goals. Furthermore, the economic capacity of the private sector should be considered as they form part of the sectors that help in ensuring the success of a particular policy objective of a country.

The other thing is the balancing of the economic and regulatory instruments, until date, such instruments have relied on governments when considering water pollution in such countries. About economic instruments such as charging fees and fines to water, polluters have been introduced but to a minor extent especially in countries that are industrialized (Afshar et al., 2015). Unlike economic instruments, regulatory instruments offer a reasonable degree of predictability on what needs to be done to reduce water pollution. The only disadvantage of the regulator approach is about its economic efficiency, given the fact that economic instruments tend to provide an incentive to individuals to take care of the environment and at the same time provide financial revenue to the particular country.

Application of water pollution control at the lowest appropriate level. The appropriate level can be described which have significant impacts. Taking for instance if significant impacts are at the community level, then management at community level should be carried out for effective results. If the neighboring community is the one that is significantly affected, then the management level should be somewhere higher than the community level. On a larger scope, the management can be at national level since the impacts of management at this level can benefit all the individuals at lower levels. This means that all the members of the country will benefit from a particular implementation. When the impacts are significantly high for many nations, then the management should be at international level.

Participatory approach should be encouraged where all the relevant stakeholders are involved in the process. This principle is the best, especially when trying to promote awareness about the importance of water pollution control. In this case, the decisions are made by considering the opinions from the public, and public should be allowed to share their views. In addition to this, the public should be provided with open access information related to water pollution. Through this, individuals will be able to understand what they are doing wrong and what needs to be done to protect the environment.

10.4. OTHER BOARDS INVOLVED

In conjunction with WHO and UNEP, other bodies have been involved in the formulation and implementation of the above principles. They include the United Nations Centre for Human Settlements (UNCHS), the Food and Agriculture Organization of the United Nations (FAO) and experts from bilateral agencies who are members of the Water Supply and Sanitation Collaborative Council (Alamanos et al., 2018).

United Nations Centre for Human Settlements (UNCHS) is a body within the United Nations that looks at the settlements of the public. Its mandate is to ensure that the public lives in a good environment that will not their health. Water pollution is normally seen to affect the settlements of individuals especially those who live close to the water bodies. Ideally, there are pollutants that settle under the water body and rise the water levels which force the habitant to displaces. In addition to this, the individuals that live in this particular area are prone to diseases. This body works to ensure protection of humans from such conditions.

Food and Agriculture Organization of the United Nations (FAO) is a specialized agency of the United Nations that are leading the international efforts to defeat hunger. The main objective of FAO is to achieve food security and nutrition. Water forms part of food and is an essential component of cooking. As such, the involvement of the FAO it is to ensure that individuals get access to food as per their need.

Figure 10.11. Food and Agriculture Organization ensures food nutrition and security.

Source: www.fao.org.

Another body is the International Waste Management Association, which is tasked with the role of pioneering scientific assessment globally on the state of waste management and in conjunction with the same calls for action to the international community. In conjunction with UNEP, they were able to create a document that establishes a rationale for the wastewater management tools (Araujo et al., 2015). The tool considers that waste and resource management along with climate change mitigation is major contributor to sustainable development. The document provides a major outlook on the possible way forward on waste management in which it draws conclusions and makes recommendations regarding the same. To add on the sustainable development goals, this book provides an outlook that sets forth Global Waste Management Goals and a Global Call to Action to achieve those goals.

All the above-mentioned bodies are international but most countries have their own means of managing water pollution. However, even with independent bodies, countries all around the globe need the above-stated agencies to achieve the water management a success.

BIBLIOGRAPHY

1. Afroz, R., Masud, M. M., Akhtar, R., & Duasa, J. B. (2014). Water pollution: Challenges and future direction for water resource management policies in Malaysia. *Environment and Urbanization ASIA*, *5*(1), 63–81.
2. Afshar, A., Massoumi, F., Afshar, A., & Mariño, M. A. (2015). State of the art review of ant colony optimization applications in water resource management. *Water Resources Management*, *29*(11), 3891–3904.
3. Alamanos, A., Mylopoulos, N., Loukas, A., & Gaitanaros, D. (2018). An integrated multicriteria analysis tool for evaluating water resource management strategies. *Water*, *10*(12), 1795.
4. Ali, R., Kuriqi, A., Abubaker, S., & Kisi, O. (2019). Hydrologic alteration at the upper and middle part of the Yangtze river, China: towards sustainable water resource management under increasing water exploitation. *Sustainability*, *11*(19), 5176.
5. Altenburger, R., Ait-Aissa, S., Antczak, P., Backhaus, T., Barceló, D., Seiler, T. B.,... & Brack, W. (2015). Future water quality monitoring—adapting tools to deal with mixtures of pollutants in water resource management. *Science of the Total Environment*, *512*, 540–551.
6. Araujo, R. S., da Gloria Alves, M., de Melo, M. T. C., Chrispim, Z. M., Mendes, M. P., & Júnior, G. C. S. (2015). Water resource management: A comparative evaluation of Brazil, Rio de Janeiro, the European Union, and Portugal. *Science of the Total Environment*, *511*, 815–828.
7. Ashton, P., & Turton, A. (2009). Water and security in sub-Saharan Africa: Emerging concepts and their implications for effective water resource management in the southern African region. In *Facing Global Environmental Change* (pp. 661–674). Springer, Berlin, Heidelberg.

8. Badham, J., Elsawah, S., Guillaume, J. H., Hamilton, S. H., Hunt, R. J., Jakeman, A. J.,... & Bammer, G. (2019). Effective modeling for Integrated Water Resource Management: A guide to contextual practices by phases and steps and future opportunities. *Environmental Modelling & Software*, *116*, 40–56.
9. Bagatin, R., Klemeš, J. J., Reverberi, A. P., & Huisingh, D. (2014). Conservation and improvements in water resource management: a global challenge. *Journal of Cleaner Production*, *77*, 1–9.
10. Bouwer, H. (2000). Integrated water management: emerging issues and challenges. *Agricultural Water Management*, *45*(3), 217–228.
11. Bouwer, H. (2002). Integrated water management for the 21st century: problems and solutions. *Journal of Irrigation and Drainage Engineering*, *128*(4), 193–202.
12. Brack, W., Altenburger, R., Schüürmann, G., Krauss, M., Herráez, D. L., van Gils, J.,... & de Aragão Umbuzeiro, G. (2015). The SOLUTIONS project: challenges and responses for present and future emerging pollutants in land and water resources management. *Science of the Total Environment*, *503*, 22–31.
13. Butt, M. J., & Bilal, M. (2011). Application of snowmelt runoff model for water resource management. *Hydrological Processes*, *25*(24), 3735–3747.
14. Cabrera, E., & Custodio, E. (2014). Urban water, an essential part of Integrated Water Resources Management. *Integrated Water Resources Management in the 21st Century: Revisiting the Paradigm*, 169.
15. Carr, G., Blöschl, G., & Loucks, D. P. (2012). Evaluating participation in water resource management: A review. *Water Resources Research*, *48*(11).
16. Cook, B. R., & Spray, C. J. (2012). Ecosystem services and integrated water resource management: Different paths to the same end? *Journal of Environmental Management*, *109*, 93–100.
17. Dobriyal, P., Qureshi, A., Badola, R., & Hussain, S. A. (2012). A review of the methods available for estimating soil moisture and its implications for water resource management. *Journal of Hydrology*, *458*, 110–117.
18. El Bedawy, R. (2014). Water resources management: alarming crisis for Egypt. *J. Mgmt. & Sustainability*, *4*, 108.

19. Fang, S., Xu, L., Pei, H., Liu, Y., Liu, Z., Zhu, Y.,... & Zhang, H. (2013). An integrated approach to snowmelt flood forecasting in water resource management. *IEEE Transactions on Industrial Informatics, 10*(1), 548–558.
20. Figuères, C., Rockström, J., & Tortajada, C. (Eds.). (2003). *Rethinking Water Management: Innovative Approaches to Contemporary Issues*. Routledge.
21. Fuhrmann, P. (2012, October). Micro pollutants and their role in the implementation of water management. In *8th EWA Brussels Conference* (Vol. 30).
22. Gao, H., Wei, T., Lou, I., Yang, Z., Shen, Z., & Li, Y. (2014). Water saving effect on integrated water resource management. *Resources, Conservation and Recycling, 93*, 50–58.
23. Geissen, V., Mol, H., Klumpp, E., Umlauf, G., Nadal, M., Van der Ploeg, M.,... & Ritsema, C. J. (2015). Emerging pollutants in the environment: a challenge for water resource management. *International Soil and Water Conservation Research, 3*(1), 57–65.
24. Ghosh, N. C. (2017). Integrated Water Resources Management. *Sustainable Utilization of Natural Resources*, 459.
25. Grizzetti, B., Lanzanova, D., Liquete, C., Reynaud, A., & Cardoso, A. C. (2016). Assessing water ecosystem services for water resource management. *Environmental Science & Policy, 61*, 194–203.
26. Hirji, R., & Ibrekk, H. O. (2001). *Environmental and Water Resources Management*. World Bank, Environment Department.
27. Jackson, S. (2006). Compartmentalising culture: the articulation and consideration of Indigenous values in water resource management. *Australian Geographer, 37*(1), 19–31.
28. Kim, J. H., Keane, T. D., & Bernard, E. A. (2015). Fragmented local governance and water resource management outcomes. *Journal of Environmental Management, 150*, 378–386.
29. Kirschke, S., & Newig, J. (2021). Complexity in Water Management and Governance. In *Handbook of Water Resources Management: Discourses, Concepts and Examples* (pp. 801–810). Springer, Cham.
30. Krampe, F. (2017). Water for peace? Post-conflict water resource management in Kosovo. *Cooperation and Conflict, 52*(2), 147–165.
31. Larson, K. L., & Lach, D. (2008). Participants and non-participants of place-based groups: An assessment of attitudes and implications for public participation in water resource management. *Journal of Environmental Management, 88*(4), 817–830.

32. Leahy, J. E., & Anderson, D. H. (2008). Trust factors in community–water resource management agency relationships. *Landscape and Urban Planning*, *87*(2), 100–107.
33. Lopez, A., Fung, F., New, M., Watts, G., Weston, A., & Wilby, R. L. (2009). From climate model ensembles to climate change impacts and adaptation: A case study of water resource management in the southwest of England. *Water Resources Research*, *45*(8).
34. Ludwig, F., van Slobbe, E., & Cofino, W. (2014). Climate change adaptation and Integrated Water Resource Management in the water sector. *Journal of Hydrology*, *518*, 235–242.
35. Muller, M. (2015). The'nexus' as a step back towards a more coherent water resource management paradigm. *Water Alternatives*, *8*(1).
36. Nazemi, A., & Wheater, H. S. (2015). On inclusion of water resource management in Earth system models–Part 1: Problem definition and representation of water demand. *Hydrology and Earth System Sciences*, *19*(1), 33–61.
37. Olmstead, S. M. (2014). Climate change adaptation and water resource management: A review of the literature. *Energy Economics*, *46*, 500–509.
38. Olsson, J. A., & Andersson, L. (2006). Possibilities and problems with the use of models as a communication tool in water resource management. In *Integrated Assessment of Water Resources and Global Change* (pp. 97–110). Springer, Dordrecht.
39. Owusu, P. A., Asumadu-Sarkodie, S., & Ameyo, P. (2016). A review of Ghana's water resource management and the future prospect. *Cogent Engineering*, *3*(1), 1164275.
40. Partha, S., Desai, A. V., Sumanta, L., & Ghosh, S. K. (2019). Advanced porous materials for sensing, capture and detoxification of organic pollutants toward water remediation. *ACS Sustainable Chemistry & Engineering*, *7*(8), 7456–7478.
41. Prinz, D., & Singh, A. K. (2000). Water resources in arid regions and their sustainable management. *Annals of Arid Zone*, *39*(3), 251–272.
42. Singh, A. K., Sathya, M., Verma, S., Kumar, A., & Jayakumar, S. (2021). Assessment of Anthropogenic Pressure and Population Attitude for the Conservation of Kanwar Wetland, Begusarai, India: A Case Study. *Pollutants and Water Management: Resources, Strategies and Scarcity*, 22–46.

43. Singh, J., Yadav, P., Pal, A. K., & Mishra, V. (2020). Water pollutants: Origin and status. In *Sensors in Water Pollutants Monitoring: Role of Material* (pp. 5–20). Springer, Singapore.
44. Singh, P., Singh, R., Singh, V. K., & Bhadouria, R. (2021). Pollutants and Water Management.
45. Sroufe, R. P. (2018). Integrated Management Resources Guide. In *Integrated Management*. Emerald Publishing Limited.
46. van Koppen, B., & Schreiner, B. (2014). Moving beyond integrated water resource management: developmental water management in South Africa. *International Journal of Water Resources Development, 30*(3), 543–558.
47. Vyas, J. N., & Nath, S. (2021). The Role of Government and the Public in Water Resource Management in India. *Pollutants and Water Management: Resources, Strategies and Scarcity*, 399–415.
48. Zhao, Z. Y., Zuo, J., & Zillante, G. (2017). Transformation of water resource management: a case study of the South-to-North Water Diversion project. *Journal of Cleaner Production, 163*, 136–145.

INDEX

F

G

H

I

L

N

O

P

R

S